Avoiding Static Ignition Hazards in Chemical Operations

This is one in a series of CCPS *Concept* books. A complete list of publications available from CCPS appears at the end of this book.

Avoiding Static Ignition Hazards in Chemical Operations

A CCPS CONCEPT BOOK

LAURENCE G. BRITTON

AMERICAN INSTITUTE OF CHEMICAL ENGINEERS

CENTER FOR CHEMICAL PROCESS SAFETY

CENTER FOR CHEMICAL PROCESS SAFETY
of the
American Institute of Chemical Engineers
3 Park Avenue, New York, New York 10016-5901 USA

Library of Congress Cataloging-in Publication Data
Britton, Laurence G., 1951–
Avoiding static ignition hazards in chemical operations / Laurence G. Britton.
p. cm. — (A CCPS concept book)
Includes bibliographical references and index.
ISBN 0-8169-0800-1
1. Electrostatics. 2. Chemical plants—Safety measures.
I. Title. II. Series
TP155.5.B75 1999 99-35855
660´.2804—dc21

CONTENTS

3

EVALUATING THE HAZARD OF STATIC ELECTRICITY

4

CONTROLLING ELECTROSTATIC HAZARDS

5

FLAMMABLE LIQUIDS, VAPORS, AND GASES

6

POWDERS AND SOLIDS

PREFACE

For nearly 40 years the American Institute of Chemical Engineers (AIChE) has been involved with process safety and loss control issues in the chemical, petrochemical, hydrocarbon process and related industries and facilities. AIChE publications are information resources for the chemical engineering and other professions on the causes of process incidents and the means of preventing their occurrence and mitigating their consequences.

The Center for Chemical Process Safety, a Directorate of the AIChE, was established in 1985 to develop and disseminate information for use in promoting the safe operation of chemical facilities and the prevention of chemical process incidents. With the support and direction of its Advisory and Management Boards, CCPS established a multifaceted program to address the need for process safety technology and management systems to reduce potential exposures to the public, the environment, personnel and facilities. This program entails the development, publication and dissemination of Guidelines relating to specific areas of process safety; organizing, convening and conducting seminars, symposia, training programs and meetings on process safety-related matters; and cooperating with other organizations and institutions, internationally and domestically, to promote process safety. Within the past several years, CCPS extended its publication program to include a "Concept Series" of books. These books are focused on more specific topics than books in the "Guideline Series" and are intended to complement them. With the issuance of this book, CCPS will have published over 40 books.

CCPS activities are supported by the funding and technical expertise of over 90 corporations. Several government agencies and nonprofit and academic institutions participate in CCPS endeavors.

In 1989, CCPS published the landmark *Guidelines for the Technical Management of Chemical Process Safety*. This book presented a model for Process Safety Management built on twelve distinct, essential and interrelated elements. The foreword to that book stated:

> For the first time, all the essential elements and components of a model of a technical management program in chemical process safety have been assembled in one document. We believe the Guidelines provide the umbrella under which all other CCPS Technical Guidelines will be promulgated.

This "Concept Series" book supports several of the twelve elements of process safety enunciated in *Guidelines for the Technical Management of Chemical Process Safety*, including process knowledge and documentation, process risk management and enhancement of process safety knowledge. The purpose of the book is to assist designers and operators of chemical facilities to understand and reduce potential hazards associated with static electricity.

ACKNOWLEDGMENTS

The American Institute of Chemical Engineers and the Center for Chemical Process Safety (CCPS) express their gratitude to Union Carbide Corporation for their generous support in the preparation of this Concept Book. This book was written by Dr. Laurence G. Britton of Union Carbide Corporation (UCC). The author earned a BSc with 1st Class Honors and a PhD in Fuel and Combustion Science at the University of Leeds, England. In January 1978 he was appointed Research Fellow in Prof. A.W. "Bill" Bright's Applied Electrostatics Group at Southampton University where he worked primarily under military contract on various static ignition topics. In mid-1981 he joined Union Carbide's Safety Research Laboratory as Senior Combustion Scientist, later holding posts of Project Scientist and Research Scientist. He is currently a Principal Engineer in UCC's Process Safety Technology Group, involved in process fire and explosion hazards research. He is a Fellow of the Institute of Energy, a Member of AIChE and the Combustion Institute, and is a Chartered Engineer in the UK. He has served on NFPA's Technical Committees for "Static Electricity" (1983–), "Explosion Protection Systems" (1983–), and "Classification and Properties of Hazardous Chemicals" (1987–), the CCPS "Engineering Practices" Subcommittee (1988–) and more recently ASTM's "Hazard Potential of Chemicals" Committee plus several ad hoc committees such as CMA's "Flame Resistant Clothing Issues Group". He has authored numerous technical articles and book contributions and twice received AIChE's "Bill Doyle" Award for best loss prevention paper.

CCPS also acknowledges the contributions of other Engineering Practices Subcommittee members in recognizing the need for a Concept Book on this subject that would bridge the gap between various existing codes or standards and the general literature. The Chair of the Engineering Practices Subcommittee is Mr. Robert Walz of ABB Lummus Global Inc. Other members were: Messrs. Stephen Cloutier (UOP, retired), Glenn Davis (DuPont),

Dennis Hendershot (Rohm & Haas Company), Kenneth Linder (HSB Industrial Risk Insurers), Peter N. Lodal (Eastman Chemical Company), John Noronha (Eastman Kodak), Carl Schiappa (Dow Chemical) and William Thornberg (HSB Industrial Risk Insurers). Dr. Martin E. Gluckstein, P.E., was the CCPS Staff Liaison and was responsible for the overall administration and coordination of the project.

Before publication, all CCPS books are subjected to a thorough peer review process. CCPS also gratefully acknowledges the thoughtful comments and suggestions of the peer reviewers: Walter L. Frank (EQE International), Thomas O. Gibson (AIChE/CCPS), Gregory Keeports (Rohm & Haas Company), Joe Kulak (Novartis Corporation), Peter N. Lodal (Eastman Chemical Company), Dr. Robert A. Mancini (BP-Amoco), Dr. Thomas H. Pratt (Burgoyne Associates), James Reppermund (DuPont), Richard F. Schwab (Allied Signal, retired), Walter Silowka (Air Products and Chemicals Inc) and Dr. Jan C. Windhorst (Nova Chemicals Ltd).

Plates 1, 2, 4, and 5 were kindly provided by Dr. Martin Glor, Institute of Safety & Security, Basle, Switzerland.

DISCLAIMER

This work expresses only the opinions of the author and does not necessarily represent the opinions or practices either of Union Carbide Corporation or any other single company. Unpublished case histories used to illustrate the text were in most cases kindly supplied by peers from other companies and are used without either reference or identification of the company involved; this should not imply that the accounts are undocumented. The book should be used in conjunction with published guidelines and recommended practices such as NFPA 77 and references [1-7].

I

INTRODUCTION

1-1. Purpose

The purpose of this book is to assist the user in controlling the hazards associated with the generation, accumulation, and discharge of static electricity by presenting

- a basic understanding of the nature of static electricity,
- methods for identifying and assessing the hazards of static electricity,
- relevant practical aspects of ignition and combustion,
- techniques for controlling the hazards of static electricity,
- previously unpublished information and illustrative case histories,
- an extensive literature review, and
- appendix material including data tabulations and glossary of terms used in the text.

A number of introductory "static hazard" texts have been published [25, 33, 67, 72, 73]. Useful practical guidelines have also been published by ESCIS [4] and Walmsley [23], respectively based on the contemporary internal practices of Ciba-Geigy [196] and the Royal Dutch/Shell Group [189]. General reading, including electrostatic applications, can be found in the textbooks by Moore [109] and Cross [110].

In 1964, B. F. Skinner observed that "Education is what survives when what has been learnt has been forgotten" [219]. Another survivor is information appearing in such documents as company practices and published codes. Recommendations of uncertain origin are in some cases copied from previous editions or from other documents without thorough re-evaluation.

Since derivations and literature references are typically not given in detail (often not at all) it is difficult for users to recognize errors or assess the validity of some statements when applied to a particular situation. It would be helpful for the user to appreciate limitations of applicability and to consider risk tolerance when adopting a specific practice based on "minimum safe practices" given in codes.

The book is aimed at chemical engineers and other technically trained persons either curious about the origin of published recommended practices on static electricity or faced with one of the many situations not specifically covered. It is intended not as an introductory text but rather to help bridge the gap between recommended practices and the general literature on this subject. Owing to the wide variety of opinions expressed both in published recommended practices and in the general literature the author has carried out a critical review. In addition to material taken directly from the literature, the book contains a large amount of original information either derived from the critical review or taken from over 20 years of personal experience. The coverage should find particular application in safety analyses and accident investigations.

While only elementary electrostatic theory is typically used in this book, more advanced concepts are introduced where necessary. Examples include background discussions of tank filling and tank cleaning models. Electrostatic equations should look familiar to engineers with a background in heat transfer. This is because there is a close analogy between potential difference, governing the flow of charge, and temperature difference, governing the flow of heat. The reverse analogy has been widely used to help solve complex heat transfer problems with reference to analog resistive circuits in which temperature differences and heat flows are replaced by potential differences and currents. Table 1.1 lists some analogous quantities and equations.

In the simplest case of one-dimensional steady flow in the x direction, there is a parallel between Fourier's law for heat flow rate and Ohm's law for charge flow rate (i.e., electrical current). For three-dimensional steady-state, potential and temperature distributions are both governed by Laplace's equation. The right-hand terms in Poisson's equation are (Q_V/ε) = (volumetric charge density/permittivity) and (Q_G/k) = (volumetric heat generation rate/thermal conductivity). The respective units of these terms are (V m^{-2}) and (K m^{-2}). Representations of isopotential and isothermal surfaces are known respectively as potential or temperature fields. Lines of constant potential gradient ("electric field lines") normal to isopotential surfaces are similar to lines of constant temperature gradient ("lines of flow") normal to

TABLE 1.1 *Heat Transfer Analogy*

Electricity	Heat
ϕ = potential (V)	T = temperature (K)
V = potential difference (V)	ΔT = temperature difference (K)
Q = charge (C)	W = heat energy (J)
I = current (C s^{-1})	Q = heat flow rate (J s^{-1})
C = capacitance (C V^{-1})	C = thermal capacity (J K^{-1})
J = current density (C s^{-1} m^{-2})	q = heat flux (J s^{-1} m^{-2})
κ = electrical conductivity (C m^{-1} s^{-1} V^{-1})	k = thermal conductivity (J m^{-1} s^{-1} K^{-1})
Ohm's law (κ constant) $I = \kappa AV/x$	Fourier's law (k constant) $Q = kA\Delta T/x$
Laplace's equation $\delta^2\phi/\delta x^2 + \delta^2\phi/\delta y^2 + \delta^2\phi/\delta z^2 = 0$	Laplace's equation $\delta^2T/\delta x^2 + \delta^2T/\delta y^2 + \delta^2T/\delta z^2 = 0$
Poisson's equation $\delta^2\phi/\delta x^2 + \delta^2\phi/\delta y^2 + \delta^2\phi/\delta z^2 = -Q_V/\varepsilon$	Poisson's equation $\delta^2T/\delta x^2 + \delta^2T/\delta y^2 + \delta^2T/\delta z^2 = -Q_G/k$

isothermal surfaces. In electrostatic books, isopotentials are often described as "equipotentials."

To facilitate solution of the differential equations, analogous simplifying assumptions can be made. For example, owing to the high thermal conductivity of steel, the internal temperature of a hot steel billet upon being plunged into a large reservoir of cold water might be assumed spatially constant. The water reservoir might be considered as a "heat sink" at constant temperature into which heat is transferred across the boundary. Similarly, the potential of charged, electrically conductive liquid poured into a poorly conducting plastic jug might be assumed spatially constant. The large surrounding "charge sink" is the earth into which charge is transferred across the jug wall. The solution to the steel billet problem [209] can be written in terms of temperature difference between the steel and the water as

$$\Delta T = \Delta T_0 \exp(-B_i F_0) \tag{1-1}$$

where ΔT_0 is the initial temperature difference and B_i, F_0 are respectively the Biot and Fourier dimensionless groups. If time is extracted from the Fourier number, Eq. (1-1) can be rewritten

$$\Delta T = \Delta T_0 \exp(-t/\tau) \tag{1-2}$$

where τ is a time constant comprising the remaining terms of the dimensionless groups. Since heat or heat density is proportional to temperature difference, this result is analogous to Eq. (2-3.7) governing the rate of charge decay.

The commonest assumption when solving Poisson's equation for potential distribution in charged liquids and powders is that the volumetric charge density is spatially constant. The assumption allows analytical solution in simple geometries such as spheres (Appendix C) and simplifies computer solution in more complex geometry via finite difference or element techniques. It is analogous to the heat transfer assumption that the volumetric heat generation rate is not only independent of time but also of position (for example, a dispersed radioisotope). Models based on the assumption of spatially constant charge density form the basis for setting flow rate limits during tank loading. The assumption implies that the filling pipe in a tank releases charge at a uniform rate throughout the liquid space and the tank is always completely full of liquid. As illustrated in the text by several case histories, such models can fail in practice owing to their simplifying assumptions. For example, poorly conducting froth layers might buoy up conductive solids such as chunks of ice or hydrates derived from the tank or inlet line, allowing static discharges to occur during transfer of relatively conductive liquids such as esters. Static discharges from nonconductive liquids are encouraged by inlet pipe orientations that allow convection of excessive charge to the free surface.

While it is important to appreciate mechanisms for production of static discharges it is often more important to recognize the factors influencing flammability and ease of ignition. As illustrated in the text by several case histories, static ignition can sometimes occur under conditions involving an unrecognized flammability hazard, such as a liquid at less than its flash point or a tank equipped with nitrogen padding. As discussed in Chapter 5, a side-effect of splash filling a tank is the associated dilution of vapor near the liquid surface via air entrainment. Since ignition via brush discharges typically occurs within 10 cm of the surface, air dilution might create an ignition hazard above liquids that would otherwise generate local vapor concentrations above their UFLs. The "ignition" component of "static ignition hazards" represents most of the problem suggested by the title of this book and consequently a large part of the text is concerned with flammability. In this respect the book differs significantly from other texts on static hazards.

1-2. Exclusions

This book is focused on static ignition hazards in chemical operations. The following are not specifically covered

- *Intrinsic Safety.* Static electrical concepts such as minimum ignition energy do not directly apply when assessing the safety of electrical circuits such as radios, flashlights and instrumentation. Intrinsically safe electrical equipment is usually available which has been subjected to fault analysis and testing. The equipment must be certified for the flammable atmosphere in which it will be used (NFPA 497). Refer to texts on Intrinsic Safety such as [63].
- *Stray Electrical Currents and Induced Radio Frequency Currents.* For information on stray currents see API 2003 [3]. For information on both hazards see the author's review "Sources of Ignition" in [157].
- *Electrostatic Discharge (ESD) Damage to Electronic Equipment.*
- *Marine Tankers and Barges* (see [5] "ISGOTT").
- *Web and Printing Processes* (see NFPA 77).
- *Lightning Protection* (see NFPA 780).

1-3. Units

The SI system of units is normally used. For descriptions in the text, units were selected either to be most familiar to the user or most appropriate for the application. For example, because the scale is nonlinear, breakdown strength is expressed in length units of the order 1 cm or less, reflecting that measured values do not apply over one meter. Additionally, since the thickness of insulating layers is often measured in mils, the mil equivalent is given to avoid confusion between μm (micron), mil, and millimeter. For ease of reading, mixed SI and cgs units are sometimes used to avoid introducing exponents or small decimals; an example is the expression of dust concentration in g/m^3 rather than kg/m^3. Volumes and flow rates are expressed in terms of US gallons where this is helpful to the user. Where conversions are made between SI and other unit systems, values are often rounded off where lack of precision does not justify exact numbers.

1-4. Organization of the Book

The practical content of this book is mostly contained in Chapter 5 (gases and liquids) and Chapter 6 (powders and hybrid mixtures) with other chapters providing supporting material. Chapter 2 contains a brief explanation of the nature of static electricity followed by a detailed discussion of the characteristics and effective energies of different static discharges. Since this chapter is relatively technical the reader might wish to refer back to it as needed. Chapter 3 discusses techniques for evaluating static hazards including the instrumentation that is frequently used. Chapter 4 describes general bonding, grounding and other techniques used either to control static or prevent ignition. Following Chapter 6 are a series of appendices containing additional explanatory material, data tables, mathematical relationships, references, and glossary. The book is extensively cross-referenced by chapter and section. Complete text references are given in the "References" section, except for documents issued by ASTM or by the National Fire Protection Association (1, Batterymarch Park, Quincy, MA 02269-9101, USA) which are referred to in the text by their document numbers (e.g., ASTM E 502, NFPA 77).

2

FUNDAMENTALS OF STATIC ELECTRICITY

2-1. What Is Static Electricity?

As the name implies, the term originally referred to physical phenomena associated with charges at rest, such as on charged, isolated conductors. However, as used in this book, "static" charges may be either be at rest or moving. The elementary unit of negative charge is the electron, which carries -1.6×10^{-19} Coulombs of charge. A positive charge is equivalent to the absence of electrons. In semiconductor theory a group of covalent bonds deficient of one electron is treated as a mobile positively charged entity, or "hole." This concept is used to describe the properties of semiconductive crystals used in transistors. However in the context of this book it is best to think of the flow of positive charge as a flow of positively charged particles or ions.

Static electricity hazards or nuisances arise when charge separation occurs leading to an accumulation of one sign of charge within some defined boundary, such as inside a container. The work performed in separating the charges results in differences of potential within or across the defined boundary and the accompanying generation of electric fields. If an electric field locally exceeds some threshold value, electrical breakdown of the intervening medium occurs in the form of a static discharge. This might come as a shock.

The "tingling" effects of static are caused by mutual repulsion between strands of hair carrying the same sign of charge, which tends to make them stand up. The phenomena occur either as the result of polarization (2-2.1) or a net charge on the body. When the body is polarized by a strong electric field, the charged strands of hair are both repelled from one another and attracted in the direction of the electric field. This can be especially hair-raising.

2-1.1. Charge Separation

This occurs in a variety of ways. When solid surfaces are placed in contact, an electronic rearrangement occurs to minimize the energy at the interface. Since this process is generally not reversible, charge separation occurs when contact between the surfaces is lost. If the interface is disrupted at a rate faster than equilibrium conditions can be established, additional charge separation occurs. However, the maximum surface charge density is limited by electrical breakdown in the gap between the separating surfaces. During separation of nonconductive plastic sheets, one sheet gains a net positive charge and the other gains an equal quantity of negative charge. If this process occurs in air, corona discharges in the gap formed between the sheets limit the maximum surface charge density to $\leq 2.65 \times 10^{-5}$ C/m^2 (6-2.1.1). As a stream of water breaks up, ions associated with aligned water dipoles at the water–air interface separate into the fine mist created as the surface layer shears away, while ions of predominantly opposite sign separate into the coarser droplets formed from the body of the water stream. This results in a charged water mist after the coarser droplets settle out (5-6.1). Ionic charge-carrying species in liquids are adsorbed nonuniformly at the wall of a pipe such that one sign of charge predominates in a tightly held "fixed layer" while the countercharge is situated farther from the wall in a less tightly held "diffuse layer." When the liquid is pumped through the pipe the diffuse layer shears away and is convected downstream. The flow of charge is equivalent to a charging current or "streaming" current (5-2.1 and 5-3.1.1).

2-1.2. Magnitude of Current and Potential

Static electricity hazards and nuisances are typified by the generation of large potentials (0.1–100 kV) by small charging currents (0.01–100 μA) flowing in high resistance circuits (10^8–10^{15} Ω). This in part differentiates static electricity from other electrical phenomena. For example, stray currents in low resistance circuits are typically of the order 1 A for potential differences of the order 1 volt (A-4-1.3). The electric field at any point in relation to a conductor is proportional to its potential, while magnetic field is proportional to the current flowing through the conductor. Since static electricity involves high potentials and very low currents, it can be differentiated from "current electricity" phenomena by its associated electric field but the absence of any significant magnetic field.

2-1.3. Concentration of Charged Species

The occurrence of static electricity is highly dependent on the presence of charged chemical species at extremely small concentrations. This is because only a minuscule fraction of an electrostatically charged substance carries a net charge. One Coulomb represents the same charge as 6.25×10^{18} electrons, or an equal number of ionic species each carrying one elementary unit of charge. A mole of substance contains the Avogadro number, 6.023×10^{23} molecules. Hence a charge density of one Coulomb per mole is equivalent to 1 molecule in 96,400 (~10 ppm) carrying an elementary charge. One Coulomb per mole is an extremely large charge density. The volumetric charge densities found in charged liquids typically range from 1 to 5000 $\mu C/m^3$. For a typical maximum charge density of 1000 $\mu C/m^3$, assuming a liquid with specific gravity 1.0 and molecular weight 100, the involvement of molecules in the net charge carrying process is one part per trillion.

Similarly, 1 m^3 of this substance contains 6.023×10^{27} molecules. The maximum surface charge density of $\leq 2.65 \times 10^{-5}$ C/m^2 (2.1.1) corresponds to 1.6×10^{14} electrons. If the surface charge is assumed to reside in a slice of the substance 10 Angstroms thick, containing 6.023×10^{18} molecules, approximately 27 ppm of these molecules carry an elementary charge. A similar concentration (about 7 ppm) is found for the more realistic case of a plastic with molecular weight 20,000 and specific gravity 0.8, where the charge is trapped in a surface layer ~1 μm thick.

2-1.4. Importance of Trace Contaminants

The electrostatic behavior of intrinsically nonconductive substances, such as most pure thermoplastics and saturated hydrocarbons, is generally governed by chemical species regarded as "trace contaminants." These are components that are not deliberately added and which may be present at less than detectable concentrations. Since charge separation occurs at interfaces, both the magnitude and polarity of charge transfer can be determined by contaminants that are surface active. This is particularly important for nonconductive liquids, where the electrostatic behavior can be governed by contaminants present at much less than 1 ppm (2-1.3).

An unpredictable charge density increase or polarity reversal caused by a "pro-static agent" (5-3.1.2) may lead to static ignition after years of uneventful operation under ostensibly identical conditions. For example, a change from positive to negative charging of a liquid may lead to formation of an incendive "positive brush" discharge (2-6.2). Such situations may be exacerbated by a coincidental decrease of ignition energy. An example is a temperature change that affects both charging of the condensed phase and flammability of the surrounding space. Many surface active trace contaminants increase the magnitude of charging in liquids. However, at higher concentrations they can have a beneficial effect by increasing the liquid conductivity to the extent that significant charge no longer accumulates in grounded containers. Special formulations are known as "antistatic additives" (5-2.6).

Trace contaminants are also significant at charged solid surfaces, affecting both the charging process and the surface conductivity. In ambient air atmospheres their effect is often determined by interaction with adsorbed water vapor, whose dominant concentration may be sufficiently large to form a monolayer. Topical antistatic agents for solids typically rely on interaction with adsorbed water and can lose effectiveness at low relative humidity (4-2.1).

2-1.5. Hazard Evaluation

The ignition hazard analysis in 2-5 begins with evaluating whether static electricity can accumulate, with the assumption that flammability has already been addressed. However, as reflected in Chapters 3, 5, and 6, a practical safety analysis should begin by evaluating whether a flammable mixture may be present, since this determines whether or not any ignition hazard exists in the first place. The usefulness of a hazard evaluation is determined largely by the evaluation of flammability, since this governs the ignition hazard with respect to any ignition source. Order-of-magnitude estimates may in some cases be sufficient to rule out ignition via static discharges. Conversely, if a large risk exists, or in the aftermath of an explosion, order-of-magnitude estimates may be inadequate. In such situations the coverage given in this book should be especially helpful.

2-1.6. Statistics

Static ignition statistics must be used with prudence not only because of possible misdiagnosis of the cause of ignition but also because of the way the losses are often grouped together. As discussed in Chapters 5 and 6 the likelihood of ignition is related most strongly to the ignition energy of the flammable atmosphere involved. In road tanker operations, although switch loading is sometimes considered separately, accident statistics usually fail to identify fires involving pure liquids which can represent the most hazardous case (5-1.4.1). The sheer bulk of some operations having a small likelihood of ignition, such as gasoline trucking, can lead to misinterpretation of incident rate by incorporating and hiding those products with a much greater likelihood of ignition, such as toluene. Other grouping errors may involve the use of antistatic additives in certain products (5-2.6). Powder ignition is similarly much more likely for certain products, especially where a substantial mass fraction comprises fine powder or the process by its nature tends to accumulate easily ignitable dust. The methods outlined in this book should help identify those products and operations most at risk.

2-2. Charge Generation

The principal mechanisms are

a. contact and separation of solids (including frictional charging or tribocharging),
b. fragmentation of solids having nonuniform surface charge densities,
c. shear at liquid–solid, liquid–gas and two-phase liquid interfaces,
d. gravitational separation of suspended material having nonuniform size and charge,
e. induction charging, and
f. ionic charging.

Mechanisms (a) through (d) are process-specific as discussed in Chapters 5 and 6. The latter mechanisms require the presence of large electric fields and represent processes whereby charge can be transferred between systems that are electrically isolated from each other. Charge generation may be a discrete process, in which case it is associated with a transferred charge (Coulombs) resulting in a net charge density (Coulombs per unit area, volume or mass), or can continue indefinitely, in which case the rate of production of charge (Coulombs/second) is equivalent to a charging current (Amperes). An example of a discrete charging process is the electrophorus induction charging apparatus (see "Glossary") while examples of continuous charging processes include flow of liquid or powder through pipelines.

2-2.1. Induction Charging

This is the process of momentarily grounding a conductor which has been polarized by an electric field, then removing the conductor from the electric field so that it gains a net charge. Polarization is the movement of charges to new locations by the action of an electric field acting on an ungrounded conductor, or the movement of an ungrounded conductor into an existing electric field. The electric field causes electron redistribution in the conductor and the performance of work, which raises its potential. An important feature of induction charging is that an ungrounded conductive object in an electric field can achieve a high potential and become a spark hazard without any contact or charge exchange with other objects. Once a spark occurs, the conductor gains a net charge and has become "inductively charged."

2-2.1.1. Production of Two Sparks via Induction

Consider an ungrounded conductive object whose potential has been raised by an external electric field to the extent that it sparks to a neighboring grounded object. When this process occurs the potential of the ungrounded conductor is reduced approximately to ground potential by the spark, which can be considered as a conductive channel. When the spark occurs, a net charge flows to ground so as to minimize the potential energy of the ungrounded conductor in the electric field. Should the electric field be removed, the net charge now existing on the ungrounded conductor results in an elevated potential relative to ground and a second spark may occur to a grounded object.

2-2.1.2. Practical Examples

1. An uncharged person wearing nonconductive shoes is influenced by the electric field from a plastic tote bin containing charged resin. Since human bodies are conductive, charges move in the electric field leading to polarization and an increase in body potential even though no charge is transferred. If the person touches the grounded metal frame of the tote bin, a spark (static shock) occurs owing to the potential difference between the person and ground. The spark causes a net transfer of charge to the person, who may carry this net "induced charge" out of the influence of the electric field and suffer a second shock when again touching grounded metal, since once out of the influence of the electric field the charged body gains a potential difference with respect to ground.

2. Conductive liquid is transferred into a nonconductive plastic container whose external walls have been externally charged by rubbing. The electric field due to the charged plastic walls polarizes the liquid and raises its potential. The charged liquid surface may now spark to a grounded metal object inserted into the container, such as a metal dip tube. After the charge is lost from the external plastic walls the net "induced charge" now existing on the liquid may produce a second spark as the liquid is poured out. If the conductive liquid is flammable, such as a lower alcohol or ketone, ignition might occur due to either of the two sparks.

2-2.2. Ionic Charging

This includes any of a number of processes whereby charge is transferred via corona discharge. Contact ion transfer between solids is discussed separately in 6-2. Examples of deliberate ionic charging include the electrostatic precipitator, charge neutralizers such as ionizing blowers, and electrostatic crop and paint sprayers. Hazards may be produced where ungrounded conductors are charged by ionization in strong electric fields. This often involves corona discharge to conductive edges and other surfaces having a small radius of curvature.

2-3. Charge Dissipation

This is the process by which excess charge is neutralized. It occurs principally via conduction through bulk materials, determined by volume resistivity, and over their surfaces, determined by surface resistivity. However, where charging currents and resistivities are very high, such as during silo filling with nonconductive powder, an additional important charge dissipation mechanism is the electrostatic discharge. Charge loss via conduction can be evaluated using the concept of relaxation time (τ) which is a time constant depending on the volume resistivity (ρ) and dielectric constant (ε_r) according to

$$\tau = \varepsilon_r \varepsilon_0 \rho \qquad (2\text{-}3.1)$$

where ε_0 = permittivity of vacuum (8.854×10^{-12} F/m)

Since conductivity (κ) is the reciprocal of resistivity (ρ), this equation can alternatively be written

$$\tau = \varepsilon_r \varepsilon_0 / \kappa \qquad (2\text{-}3.2)$$

Knowing ε_r and ρ (or κ), the relaxation time (τ) can be estimated. The determination of relaxation time allows the rate of charge dissipation to be calculated for any process obeying Ohm's law.

Consider the leaky parallel plate capacitor shown in Figure A-4-1.3. If the capacitor is momentarily charged and allowed to discharge through resistor R_L, so that the charging current $I_C = 0$, the leakage current I_L

$$I_L = -dQ/dt = V/R_L \tag{2-3.3}$$

By definition, the resistivity and conductivity of a conductor of length d and cross-sectional area A

$$R_L = \rho d/A = d/\kappa A \tag{2-3.4}$$

From C-1.1 the equation for capacitance of a parallel plate capacitor (plate area A, separation d) is

$$C = Q/V = A\varepsilon_r\varepsilon_0/d$$

or

$$A/d = Q/\varepsilon_r\varepsilon_0 V \tag{2-3.5}$$

From (2-3.3) through (2-3.5)

$$dQ/dt = -Q/\tau \tag{2-3.6}$$

where

$$\tau = \varepsilon_r\varepsilon_0/\kappa$$

Integrating (2-3.6)

$$Q_t = Q_0 \exp(-t/\tau) \tag{2-3.7}$$

Ohmic charge decay processes obey a first order rate law from which the charge Q_t remaining at any time t can be expressed in terms of the initial charge Q_0 and relaxation time constant τ. Using Eqs. (2-3.4) through (2-3.5) the time constant τ can alternatively be expressed as

$$\tau = R_L C \tag{2-3.8}$$

From Eq. (2-3.6) the rate of charge loss to ground is proportional only to the quantity of charge remaining and is independent of the shape of the grounded boundary. The form of the rate law is the same as that governing radioactive decay. Just as plutonium's decay rate is independent of the surrounding geometry, charge cannot be made to decay faster by packing a relaxation tank with conductive steel mesh (5-3.5). By measuring the time taken for charge to decay to e^{-1} (36.79%) of its initial value, τ (hence ρ or κ)

can be found using Eqs. (2-3.2) and (2-3.7). Alternatively, the "half-value" time ($t_{0.5}$) taken for charge to fall to one-half its initial value is related to τ by the equation $\tau = t_{0.5} / \ln 2$. Half-value time is distinguished from radioactive half-life since, unlike the latter, it can readily be modified (4-2).

Equations (2-3.7) for Q_t and Eqs. (2-3.1), (2-3.2), and (2-3.8) for τ are used extensively in static hazard analysis. Examples include selection and use of instrumentation (3-5.3) and residence time provisions for charged liquids (5-2.4).

2-3.1. Variability of Conductivity

When a voltage is applied to a dielectric (insulator), a current passes that decays with time owing to various polarization mechanisms [133]. Conductivity is always time-dependent. This general time dependency affects conductivity measurement for nonconductive liquids, where the peak initial current is used to calculate conductivity. Test methods are given in 3-5.5 and 3-5.6. Nonohmic behavior is pronounced for nonconductive liquids in plastic tanks, whose dielectric walls further complicate the charge decay rate [206].

The conductivity of solid dielectrics is roughly independent of temperature below about 20°C but increases according to an Arrhenius function at higher temperatures as processes with different activation energies dominate [133]. In the case of liquids, the conductivity continues to fall at temperatures less than 20°C and at low ambient temperatures the conductivity is only a fraction of the value measured in the laboratory (3-5.5). The conductivity of liquids can decrease by orders of magnitude if they solidify (5-2.5.5).

The surface conductivity of solids is dependent on surface conditioning, varying with the state of oxidation, ambient humidity and contamination. The volume conductivity may depend on small concentrations of impurities or additives. This effect is especially pronounced for nonconductive liquids, where orders of magnitude increases in conductivity can be observed for part-per-million concentrations of some antistatic additives (5-2.6). Nonconductive liquids such as aliphatic hydrocarbons are intrinsically nonionic and the measured conductivity is due to minor components that act as charge carriers. Fuels such as gasoline and kerosene consist of mixtures of up to about 200 different "bulk components" comprising aliphatic, naphthenic, aromatic, and unsaturated hydrocarbons whose composition is nonspecified except in terms of fuel performance, which may include up to about 11 parametric tests for gasoline and 26 in the case of aviation fuels [147]. A change in specification, such as a low sulfur requirement, is likely to lower a variety of other components that act as charge carriers, resulting in a reduction in conductivity. However, the use of conductive bulk components, such as ethanol in gasohol, might not increase the conductivity to the extent expected owing to compensating factors in the new formulation. Depending on use, fuels may contain a variety of additives (lubricity, antioxidant, anti-knock, corrosion inhibitor, detergent, biocide, antifoam, demulsifier, dye, etc.), any of which can impact conductivity. Following formulation, the conductivity is impacted by handling which may introduce water and other "heavies" from tank bottoms, or might involve preferential removal of certain components in clay filters, filter coalescers or water layers.

2-4. Charge Accumulation

This occurs whenever the rate of charge generation exceeds that of charge dissipation. One of two results may occur for a leaky capacitor system with constant charging current: either a steady state is attained in which the rate of current dissipation balances the charging current, or a static discharge occurs before such a condition can be reached (Figure A-4-1.3). In practice, charging currents may be nonconstant and the system considered may be significantly changed during the charging process. Examples include tank filling with liquid and powder loading to a hopper, in which the effect of rising level inside the container must be included in any analysis.

The hazard of charge accumulation usually increases with increased container volume. This follows from the relations given in C-1.4. For example, if a tall, cylindrical container is filled with uniformly charged, nonconductive powder, the maximum electric field appears at the wall and is proportional to the radius of the container while the maximum potential appears on the axis and is proportional to the square of the radius. It follows that for any assumed powder charge density, the electric field eventually exceeds the breakdown field of air as container radius increases. In cases where charge dissipation occurs at a significant rate, such as liquid tank filling operations, it becomes impractical to fill very large tanks fast enough to offset the rate of charge dissipation (5-4.1). Therefore, hazardous charge accumulation is typically not seen in very large storage tanks but instead the hazard maximizes at an intermediate tank size, depending primarily on flow rate, inlet pipe diameter and liquid conductivity.

Owing to the competitive processes of charge generation and dissipation, the rate at which a container is filled is an important factor whenever significant current leakage can occur. Effects can be additive. For example, during drum filling with a nonconductive liquid, higher charge densities result from higher velocities and rates of shear at the pipe wall (5-3.1.1) while there is less time for charge inside the drum to dissipate to the walls. If the drum is poorly grounded the high flow rate and charging current are likely to result in hazardous charge accumulation on the drum. If charging currents are very high the limiting charge dissipation rate becomes due to conduction through the liquid itself, and static discharges may occur inside a properly grounded drum (5-8.1).

2-5. Ignition

In order for static electricity to be a source of ignition, four conditions must be fulfilled

- The rate of charge generation must exceed the rate of dissipation, so charge can accumulate.
- A static discharge must coincide in time and space with a flammable atmosphere.
- The effective energy of the static discharge must exceed the ignition energy of the local mixture.
- A locally ignited flame must propagate into a surrounding flammable atmosphere.

Although charge accumulation and static discharges are commonplace in some systems, such as pneumatic silo filling with powders, ignition may be prevented or delayed for many years by failure to meet all four conditions. For example, a nonuniform dust cloud's ignition energy varies by orders of magnitude with time and position, while a large range of effective energies may be manifested by static discharges, also varying with time and position. For flame propagation to result, these two random events must coincide in time and place such that a local ignition occurs, then the small flame kernel so formed must propagate into a surrounding flammable mixture without being quenched. Ignition is discussed further in Sections 3-5.4, 5-1.4, 6-1.2, and 6-5.

2-5.1. Effective Energy

Also known as "equivalent energy," this is the spark ignition energy of the least easily ignitable mixture that can be ignited by a particular ignition source, such as a brush discharge under defined geometrical conditions.

The maximum effective energy is the largest effective energy exhibited by a particular category of ignition source,such as the entire category of brush discharges. Figure 2-5 illustrates the maximum effective energies of various ignition sources including types of static discharge [10,157]. For example, bulking brush discharges are believed responsible for ignition of dusts having spark ignition energies less than that of *Lycopodium* (<20 mJ). Optimum hydrogen–air mixtures can be ignited by all types of static discharge, whereas coarse dusts (ignition energy >100 mJ) in air can be ignited only by relatively energetic sparks and propagating brush discharges. The important concept here is that static discharges such as the brush and bulking brush dissipate energy over a large volume, whereas minimal ignition corresponding to the minimum spark ignition energy takes place in a very small volume having a typical diameter of 2 mm for gases such as propane and 2–3 times this for fine dusts in air. Even optimized sparks are not 100% efficient since they lose heat to the electrodes and energy in the form of shock waves, and measured MIE values are larger than theoretical values [142].

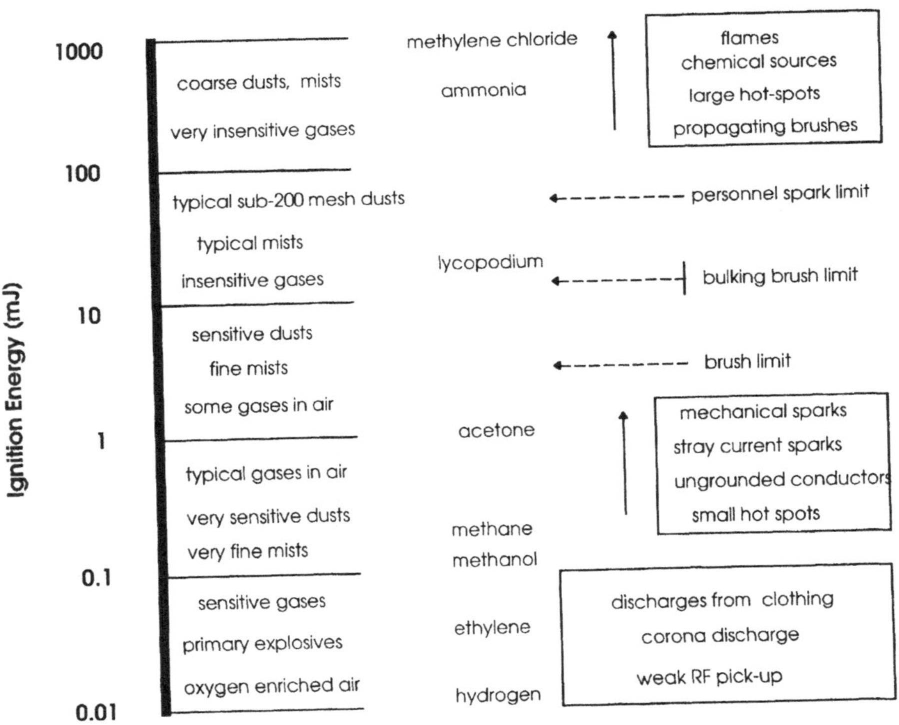

FIGURE 2-5. Ignition energies (mJ) of various materials and types of ignition source that may ignite them.

2-5.1.1. Effective Energy of Nonspark Discharges

A static discharge that is much larger in extent than the minimal ignition diameter dissipates much of its energy wastefully, and only the energy released within a local "hotspot" within the discharge contributes to ignition. Such hotspots have been observed to occur close to the metal electrode for both brush and bulking brush discharges [8,12]. The total energy dissipated in these discharges is irrelevant, although there is an apparent correlation of effective energy with the total charge transferred. This is to be expected from the $W = QV/2$ proportionality between energy W and charge Q (C-1.2).

If ignition is assumed to occur in a hotspot formed at the electrode, the local release of potential energy $W = QV/2$ is directly proportional to charge while independent of electric field except in the immediate vicinity of the electrode, since this determines the local change in potential. The electric field near the electrode becomes increasingly uniform as electrode radius increases and eventually, uniform field breakdown conditions are approached (C-2.5.3). These concepts allow first approximations for effective energy to be made. First it is assumed that air breakdown occurs at 30 kV/cm and that the electric field is approximately uniform between about 2–5 mm from the electrode (C-2.5.2). Second, it is assumed that minimal ignition occurs in a 3 mm diameter hotspot formed sufficiently far from the electrode to minimize heat losses. This distance is assumed to be about 2 mm.

A typical value of charge transferred in a positive brush discharge at a large electrode is 0.2 μC (2-6.2). The loss in $QV/2$ potential energy as this charge falls through the 3 mm distance is 0.9 mJ. The maximum energy might achieve about 10 mJ assuming a charge transfer of ≤ 1 μC, a hotspot diameter ≤ 10 mm and an average field of about 20 kV/cm. These rough estimates correctly predict the spark ignition energies of gas mixtures that have been ignited by positive brush discharges (2-6.2.1).

Similarly, a typical value of charge transferred in a bulking brush discharge is 10 μC (2-6.3) and the loss in $QV/2$ potential energy is 45 mJ. The maximum energy is 112–225 mJ for charge transfers in the range 25–50 μC. These estimates are significantly greater than the accepted maximum effective energy of about 20 mJ relative to dust ignition (2-6.3). The disparity can be explained if bulking brushes are less effective in igniting dust clouds than gas mixtures of equal spark ignition energy. Many dusts have MIEs much less than the 5–10 mJ maximum effective energy of positive brush discharges established by gas ignition tests. However, attempts to ignite sensitive dust clouds by positive brush discharges have failed (2-6.2.1). Relative to gas mixtures having the same spark ignition energies, it appears that the efficiency of dust cloud ignition by nonspark discharges is only about 10%. It is therefore possible that the maximum effective energy of the bulking brush would be closer to the estimated values if it were based on gas ignition rather than dust ignition. This implies that flammable gas in a large container should be at risk of ignition over most of its flammable range. This is consistent with the high frequency of ignition when unpurged resins are conveyed into silos containing air (6-5).

Since effective energy is an equivalent spark ignition energy, an important consideration is the maximum power density that a nonspark discharge can exert. Qualitatively, a high power density corresponds to the channeling of energy into a short-duration (50–500 ns) spark-like hotspot while a low power density corresponds to a long duration (10–250 μs) spatially diffuse discharge. These tendencies vary according to the type and polarity of the discharge. Hotspot formation close to an electrode should be encouraged by local field intensification (C-2.5.3). For nonspark discharges, local field intensification is greatest in the case of brush discharge to a small diameter electrode and smallest in the case of bulking brush discharge to a silo wall.

Even if all variables are maintained as constant as possible, nonspark discharges exhibit a distribution of maximum power densities and this introduces a probabilistic factor. The maximum effective energy attributed to any nonspark discharge category is based on a large number of repetitive ignition tests under conditions regarded as extreme. For example, in the case of the positive brush, a large area of charged surface and large diameter electrode are used. Since this might represent the top few percentiles of possible effective energies it provides a large margin of safety when applied to less severe conditions or where the discharge is a relatively isolated event.

2-6. Static Discharges

Spark, corona and brush discharge phenomena are described in detail in [137,138]. A more recent introduction is given in [139]. The inception voltage and characteristics of static discharges generally depend on the gas composition and pressure/temperature conditions. If the discharge gap is asymmetrical, the characteristics of the discharge also depend on electrode polarity. Two gas composition variables affecting electrical breakdown are ionization potential and electron affinity. For breakdown in ambient air, approximate breakdown conditions are that the average electric field in the discharge gap should exceed about 5 kV/cm and at some point the uniform breakdown field of air, about 30 kV/cm (2-6.4.1), must also be exceeded.

Discharges may be initiated by a sudden change in gas composition in an electrically stressed region. Conversely they are usually suppressed by increased gas density above atmospheric pressure (2-6.4). The effects of environmental variables on static discharge initiation are not well understood, particularly in the case of large-scale phenomena such as the bulking brush discharge, where effects due to charged powder suspensions above the bed are an added complication.

Several disparate categories of static discharge can be identified for the purposes of describing ignition and shock hazards. The most familiar (spark discharge) has an almost unlimited effective energy range. In roughly ascending order the maximum effective energies of the others are

1. Corona (positive) ≅ 0.1 mJ
2. Brush (negative) ≅ 1.0 mJ
3. Brush (positive) ≅ 10 mJ
4. Transitional Brush (author's term for energetic non-PBD discharge from nonconductive layer) ≅ 10–100 mJ
5. Bulking Brush (also known as "cone discharge") ≅ 20 mJ
6. Propagating Brush Discharge (PBD) ≅ 100–1000 mJ

The maximum effective energies given for discharge types (1–3) are based on gas ignition tests. In the case of the bulking brush the 20 mJ value reflects the ability to ignite dust clouds in air and is based on industrial loss experience. Gas mixtures of greater spark MIE might be ignited by bulking brushes as discussed in 2-5.1.1. The maximum effective energies of discharge types (4) and (6) vary widely with conditions as discussed later in this chapter. An additional type dubbed by the author a "surface streamer" (2-6.6) has frequently been mentioned in US literature under various other names but has not been subject to significant study. Another additional type is the "lightning-like" discharge which is a possibility where very large charged clouds are involved. Since lightning-like discharges have not been reported for any industrial situation they are discussed further only in the "Glossary".

2-6.1. Corona Discharge

This term includes a number of glow discharge phenomena usually occurring in highly nonuniform electric fields, either in gaps or at high voltage conductors, below the spark breakdown value (Plate 1). Coronas are often produced in the divergent field between a charged surface, such as nonconductive oil, plastic sheet or powder, and a conductor having a radius of curvature less than about 3 mm. Typical arrangements are "point–plane"

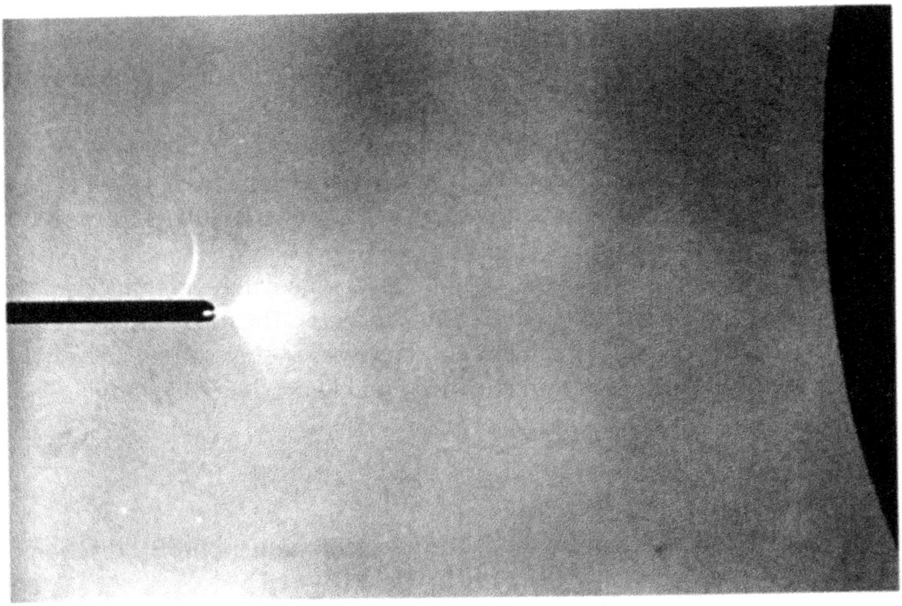

PLATE 1. Corona discharge in point–sphere gap.

and "rod–plane" geometries comprising needle or wire electrodes. The inception voltage for corona is reduced with reduced pressure and reduced radius of curvature of the electrode. Fine wires typically produce significant corona currents at about 1 kV in atmospheric air. While increased gas density usually increases the inception voltage of corona, Paschen's law (2-6.4.1) is only obeyed if changes in gap length are accompanied by changes in all other gap dimensions. Two distinct types are negative and positive point–plane coronas, where the polarity is ascribed to the electrode rather than the plane surface. With negative point the corona flows in regular "Trichel" pulses whose frequency increases with current. With positive point the current increases steadily with voltage up to about 0.1 μA after which avalanches of electrons give rise to ~1 kHz "burst corona." Pre-breakdown streamers accompanying the bursts at higher voltages may become observable as filamentary discharges more typical of brush discharges. The inception voltage for both types of corona varies with the gas composition.

Corona is usually a safe means of dissipating charge. Only very sensitive flammable mixtures such as hydrogen–air, carbon disulfide–air or fuels in oxygen enriched atmospheres (see 5-9.6) are at risk of ignition. Common fuel vapors in air can be ignited only if corona discharges have an average current in excess of 200 μA (5-9.6). This exceeds the charging currents possi-

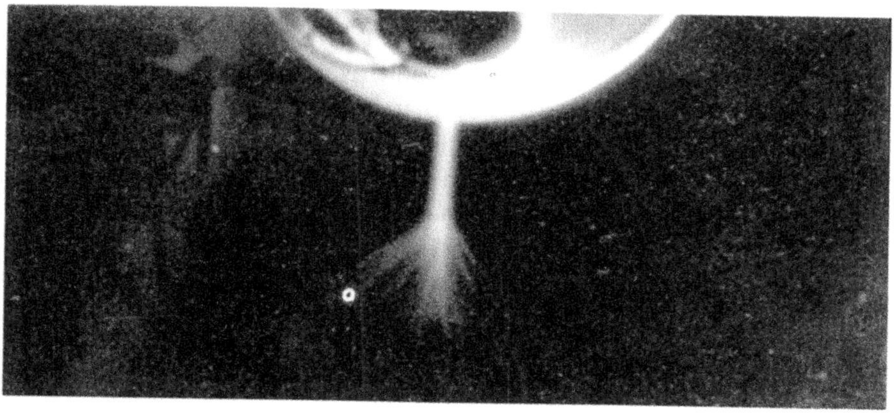

PLATE 2. Positive brush discharge from negatively charged plastic to grounded sphere.

ble in most systems. Exceptions include atmospheric electricity and some electrostatic process operations. Ignition currents for ignition sensitive gases such as acetylene in air have not been determined but might be less than 100 μA.

2-6.1.1. Positive Coronas from Active DC Neutralizers

The approach of a conductor to an active neutralizer using positive DC corona (i.e., positive needle electrode) can cause "pulsive" discharges whose peak currents attain several amperes (4-2.2.1). These have been shown to ignite stoichiometric (6.5%) ethylene in air. The discharges could be controlled by increasing the resistive coupling of the needle above 10 MΩ at 5 kV and 50 MΩ at 17 kV [39]. The phenomenon is only relevant to situations where a high voltage power supply is used.

2-6.2. Brush Discharge

Brush discharges are formed in the divergent field between a charged surface and a conductor having a radius of curvature more than about 3 mm and are a special case of point–plane coronas, requiring greater initiation voltages owing to the larger electrode radius and breakdown field. Under atmospheric conditions, brush discharges are observed at point–plane potential differences above 20–25 kV [8].

"Negative brush" discharges from positively charged surfaces appear as discrete bursts whose frequency increases with increased charging current flowing to the system. An example is liquid in a tank charged by a positive streaming current from a pipeline. Photographs show that the light-emitting region is found close to the electrode and is relatively featureless, lacking the streamers typical of positive brushes [8,71,227]. The maximum effective energy is less than 1 mJ [8,58]. The only available report of gas ignition required an unusually large (89 mm diameter) spherical electrode [35]. If the surface is instead charged negatively, "positive brush" discharges are produced. For identical geometry and charging current these occur about an order of magnitude less frequently than negative brushes while the charge transfer is about an order of magnitude greater. Typical charge transfers are 0.05–0.2 μC and maximum charge transfers are less than 1 μC. The discharges appear as a "brush" composed of streamers which converge into a single radiant channel close to the electrode (Plate 2). This channel has the highest power density within the discharge and it is here that ignition takes place [8]. Brush discharge energy increases with increased electrode diameter since the surface charge density must be greater before breakdown occurs at the electrode. Smaller electrodes are more likely to produce brush discharges but the charge transfer and effective energy are less (C-2.5.3). In work reviewed by Cross [110] the maximum charge transfer Q was found to increase with electrode diameter d according to $Q \propto d^{1.7}$. As discussed in 2-5.1.1 the effective energy of a positive brush should be roughly proportional to Q. Figure 2-6.2.1 lends some support to a power law dependence of effective energy on electrode diameter, but the curve intersections on the fuel lean side of Figure 2-6.2.1 are highly sensitive to concentration errors. Experimental errors were no doubt compounded by compiling the figure from two disparate sets of data. The result for the 15-mm electrode is least prone to error since the 0.6 mJ effective energy intercept corresponds to a less steep part of the spark MIE curve. If effective energy increases with $d^{1.7}$, predicted results are 1.4 mJ for the 25-mm and 6.3 mJ for the 60-mm electrodes, compared with 4 mJ and 8–9 mJ observed.

2-6.2.1. *Effective Energy of Brush Discharge*

This section addresses positive brushes from negatively charged, nonconductive surfaces. It does not apply to discharges from plastic surfaces having a grounded metal substrate, which are discussed in Section 2-6.2.2. Ignitions of aliphatic hydrocarbons in air have been reported for charge transfers above 0.08 μC. The tests used spherical discharge electrodes with diameters above about 7 mm [8,26]. Owing to charge conduction through the conductive flame plasma, charge transfers were deduced from previous discharges under identical conditions that did not lead to ignition. Mixtures of hexane in air were ignited by brush discharges from spherical electrodes of different diameters (15, 18, 20, 25, 35, and 60 mm) suspended above charged plastic sheet with no metal substrate [36]. The relative frequency of ignition was determined for each electrode with respect to hexane concentration. The

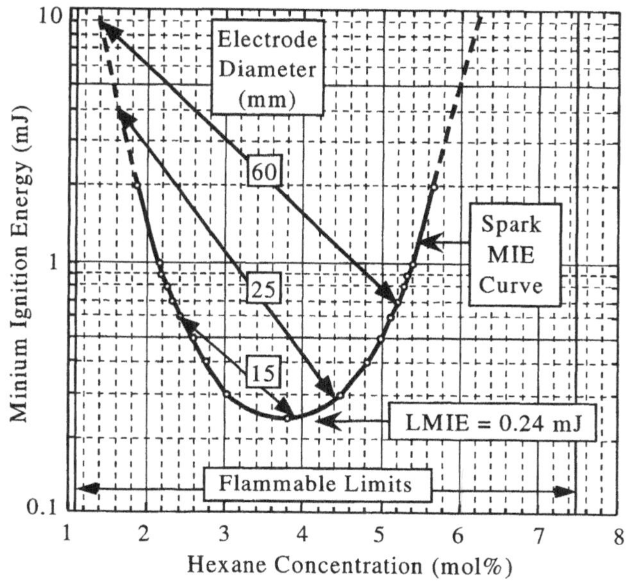

FIGURE 2-6.2.1. Ignition of hexane in air by positive brush discharges and sparks.

range of concentrations susceptible to ignition was determined by extrapolating the ignition frequency to zero. It was found that the range of concentrations that could be ignited widened with increased electrode diameter. For the 15-mm electrode, hexane was ignitable between 2.4 and 3.9% in air. The ignitable range increased to 1.6–4.4% with the 25-mm electrode and 1.4–5.2% with the 60-mm electrode.

The partial results are shown in Figure 2-6.2.1. The ranges of ignitable concentrations are superimposed on a spark ignition energy curve for hexane in air calculated from data in [56]. Both data sets are assumed to be characteristic of *n*-hexane, unbiased by the presence of methylcyclopentane or hexane isomers (dimethylbutanes and methylpentanes).

Figure 2-6.2.1 shows two interesting phenomena that have not previously been recognized. First, for a given electrode the effective energy is much greater at the lean end of the ignitable concentration range than at the rich end. Second, the concentration most susceptible to ignition by brush discharges is lean compared with the lowest minimum ignition energy (LMIE) composition most easily ignited by spark. The spark LMIE concentration of hexane is 3.8% in air (Appendix B). However, 3.8% hexane is close to where the ignition frequency for the 15-mm electrode falls to zero. The composition most frequently ignited by positive brushes was 3.2% hexane in air. No ignitions were observed above 5.2% which is considerably below the upper flammable limit (UFL) of 7.5%. Conversely, ignitions using large electrodes were observed down to about 1.4% which is not far above the lower flammable limit (LFL) of 1.1%.

Consider a tank being filled with a liquid such as gasoline whose equilibrium vapor concentration exceeds the UFL. Close to the liquid surface where small brush discharges occur, the UFL will be approached or exceeded even during the early stages of filling. Ignition via small brush discharges is only probable if the vapor concentration near the surface is much less than the UFL. Hence, gasoline vapor ignition via brush discharge is improbable even if most of the vapor space is inside the flammable range. Consider now a single-component liquid such as toluene. The vapor concentration near the surface is typically close to optimum throughout tank filling and never approaches the UFL. Hence toluene vapor ignition via brush discharge is a far more probable event (5-1.4.1).

Propane–air–nitrogen ignition tests were conducted using a 70-mm-diameter spherical electrode suspended above a negatively charged plastic sheet [57]. Since gas mixtures with ignition energies up to 3.6 mJ were ignited at an ignition frequency of 40%, positive brush discharges were attributed a maximum effective energy of about 4 mJ. This work has been widely cited without reference to ignition frequency.

Figure 2-6.2.1 shows that electrodes of diameter ≤15 mm produce brush discharges with effective energies less than 1 mJ. This is partly offset by the greater likelihood of producing brush discharges with smaller diameter electrodes. As discussed in C-2.5.3 the surface charge density for breakdown in fixed geometry decreases with decreased electrode diameter and gap, so it is to be expected that smaller diameter electrodes produce discharges with smaller effective energies. Effective energies above 1 mJ are only observed for electrodes having a diameter above 15–20 mm. At small ignition frequency, the effective energy observed in lean mixtures using large electrodes exceeds 4 mJ and attains roughly 8–9 mJ at electrode diameters of 60–70 mm.

Ignition of fine sulfur dust was demonstrated by collecting the charge from brush discharges and discharging it to ground through the spark gap in a Hartmann tube [34]. Ignitions were obtained for charge transfers above about 0.25 μC, implying that brush discharges might ignite dusts in air. However, the test technique was unrealistic. By channeling the energy through a small spark gap, the power density of the original brush discharge was radically increased in the same way a focusing lens increases the power density of a light beam. No conclusion can be drawn from these tests regarding whether a thermally stable dust can be ignited in air by a brush discharge. All other dust ignition experiments have given negative results [10,12].

Brush discharges from isolated nonconductors have not been shown to ignite combustible dusts in air. Ignition of some fine dust suspensions may be possible in the presence of flammable gas below the gas LFL (6-1.3). Thus, hybrid mixtures containing flammable gas at a concentration below the LFL may be at risk. This may include powders that contain sorbed solvents above 0.2–0.5 wt% or which slowly decompose or react in storage evolving flammable gas, such as propanol evolution from aluminum propoxide in moist air (6-1.3.2). If the nonconductor has a grounded substrate, such as plastic-coated metal, more energetic discharges can be produced that might ignite dusts in air. In this book, these are described as having a continuum of effective energies beginning with the brush, then the transitional brush, and finally the propagating brush (PBD).

2-6.2.2. *Effect of Grounded Substrate*

The effective energy of a brush discharge varies with the charge transferred to the electrode (2-5.1.1) which in turn depends on the area and charge density of the nonconductive surface when the discharge occurs. For isolated nonconductors the charge density is limited by air breakdown as discussed in 6-2.1.1. However, if the charged nonconductor is in direct contact with a conductive surface, much larger charge densities can be accumulated. This is because the electric field is primarily exerted through the nonconductor to the countercharge produced in the conductor. In other words a capacitor is formed. A common example is a metal surface coated with plastic. Charge can accumulate on the exposed plastic surface until electrical breakdown occurs spontaneously through the plastic layer to ground. Where breakdown of the layer occurs a propagating brush discharge (PBD) might be produced (2-6.5).

If the surface charge density is insufficient for a spontaneous PBD to occur, the event can be initiated by the approach of a grounded metal electrode to the surface. Tests were conducted with plastic layers from 0.15 to 8.0 mm charged by corona at source voltages from 2.5 to 75 kV [37]. Using a 20-mm spherical electrode, PBDs were obtained at surface charge densities ≥250 $\mu C/m^2$, becoming fully developed at 800 $\mu C/m^2$ (2-6.2.3). Brush discharges were obtained at all smaller charge densities provided the charging voltage was at least 5 kV.

Rubbing the plastic with felt cloth produced 1–30 $\mu C/m^2$, the largest charge densities being obtained with the thinnest layers. This is only about one tenth the surface charge density required for a PBD. Charge accumulation was limited by corona discharge in the gap between the charged layer and the oppositely charged cloth. Brush discharges from negatively charged layers less than 2 mm thick could not ignite hexane in air while layers less than 0.2 mm thick could not ignite hydrogen in air [37]. The former observation has yielded the rule of thumb that layers less than 2 mm thick, charged by simple rubbing, will not produce discharges capable of igniting common solvent vapors in air. The criterion should be conservative if applied to positively charged layers. However the criterion does not apply to surfaces which are intermittently charged by ions or by particle impact, since charge accumulation via these processes is not limited by the close proximity of a countercharged cloth. Instead, large surface charge densities can accumulate over an extended period, possibly resulting in PBDs.

2-6.2.3. *Transitional Brush Discharges*

The literature commonly states that PBDs have an effective energy of the order 1000 mJ and brush discharges have effective energies less than 4 mJ. No intermediate cases are described. Industrial experience does not support such a quantum change in effective energy. If all PBDs had effective energies of the order 1000 mJ, ignition should be commonplace under all conditions that PBDs occur. Although previously unrecognized, the concept of a more gradual transition is supported by several published studies. As layer thickness and breakdown voltage are increased, a transition region should first be reached within which PBDs (2-6.5) are not fully developed and cannot readily be distinguished from brush discharges. The author has coined the term "transitional brush" to describe discharges in this proposed transition region, which can be thought of as a discharge with effective energy between that of the brush (<10 mJ) and the fully developed PBD (>100 mJ).

Full PBD development has been described in terms of the appearance of luminous, branched channels over the entire charged surface as the initial surface charge density increases from 250–800 $\mu C/m^2$ [37]. This increase in surface charge density produces a corresponding increase in potential difference across the charged layer prior to electrical breakdown. Increased charge collection by the highly branched surface discharge results in a corresponding increase of charge transfer. Fully developed PBDs typically transfer 100–500 μC of charge [29,160] although this depends both on the area of the charged surface and the extent of lateral discharge (2-6.5.3).

Tests were conducted using an approximately 30-cm-diameter, 80-μm-thick polycarbonate film with brass backing plate. A 50-mm-diameter spherical discharge electrode was used to initiate discharge of the charged film

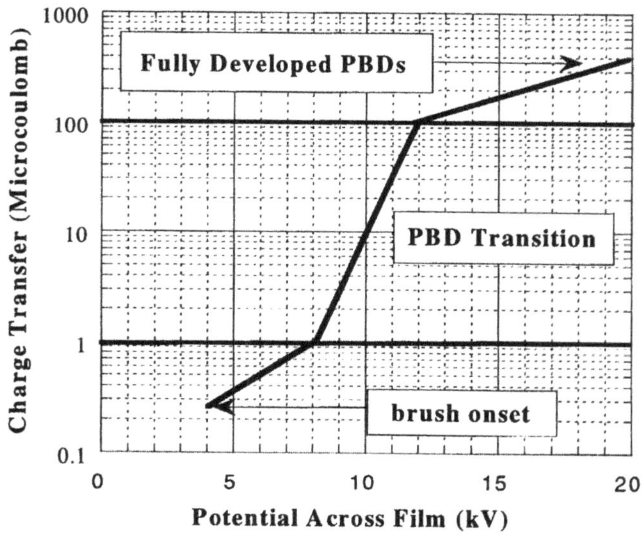

FIGURE 2-6.2.3. Transition from brush discharge to PBD on 80-μm polycarbonate film.

[29]. Adapted test results are shown schematically in Figure 2-6.2.3. No discharges were observed where the potential difference across the film was less than 4 kV, consistent with the earlier finding that brush discharges started to be produced at a corona charging voltage of 5 kV (2-6.2.2). Between 4 and 8 kV across the film, charge transfers were less than 1 μC which is consistent with large brush discharges rather than PBDs. Between 8–12 kV the charge transfer increased rapidly to 100 μC while at greater voltages the charge transfer increased at a smaller rate. This suggests that in the 8–12 kV "transition region" there was a sudden increase in charge collection efficiency from the 80-μm film as surface flashover conditions typical of PBDs were developed. The threshold for a PBD might be tentatively defined by a charge transfer of at least 10 μC. This threshold occurred at a voltage of 10 kV across the 80-μm polycarbonate film. The significance of these results is discussed further in 2-6.5.

When breakdown of the layer occurs a large voltage gradient is created over the surface to the puncture point, which becomes an effective ground, and this initiates lateral discharge to the puncture point. However, this does not explain why PBDs can propagate over much larger distances than can be achieved in air gaps at the same voltages. A proposed step-by-step mechanism is discussed in A-2-6.5. The charged double layer is conceptually modeled as a system of capacitors between and among (1) arbitrary charges on the surface and (2) their countercharges in the underlying metal. On a microscopic scale the layer surface is neither flat nor uniformly charged, so capacitance may exist between discrete charged areas of the surface. These charged areas also have some capacitance with the general surroundings. As layer thickness is increased the capacitance through the charged layer decreases, causing the electric field component across the surface to increase. Figure A-2-6.5 draws an analogy with a variable system of capacitors through the layer and across the surface; such arrangements of capacitors create a very steep gradient at the stepped discharge wavefront which allows extensive propagation for relatively small source voltages [176]. Surface flashover, which is accompanied by efficient collection of surface charge, requires a minimum surface voltage gradient of the order 4 kV/cm [176]. Transitional PBDs might be sensitive to effects that influence the proposed step-by-step propagation process, especially for thicker layers. Examples include surface topography or curvature, uniformity of surface charge density and external geometry.

2-6.2.4. Breakdown Strength of Layers

Also known as dielectric strength, this is the voltage per unit thickness at which a solid layer breaks down electrically under uniform field conditions. For a given material it depends on test parameters such as sample thickness and temperature, plus the frequency and waveform of the testing voltage. Measured values are found not to be linear with respect to thickness and higher values are found for thinner layers. Comparisons between different materials should preferably be made on samples of equal thickness under identical test conditions. Some data developed for use in capacitor design are given in Appendix B. These data indicate that a "typical" breakdown strength for cast resins such as phenol-formaldehyde is 400 volts/mil (16 volts/μm or 16 kV/mm). For polyolefins such as polyethylene a "typical" value is 1000 volts/mil (39 volts/μm or 39 kV/mm). The highest listed values correspond to polyethylene terephthalate film, averaging 4500 volts/mil (177 volts/μm or 177 kV/mm), and chloro-fluoro polymer films. The average value for PTFE film was 1500 volts/mil (59 volts/μm or 59 kV/mm) and for PCTFE film 4000 volts/mil (157 volts/μm or 157 kV/mm). A common interpretation of Figure 2-6.5 and its description in [29] is that the breakdown strength of plastic films is relatively constant at 400 volts/μm and independent of layer material, grade or thickness. This breakdown strength is 10 times the value given for polyethylene in [183] and approximately equal to the breakdown strength of many plastics when expressed in volts/mil (Appendix B). The breakdown voltages shown in Figure 2-6.5 corresponded to very thin layers less than 50 μm (2 mil). Since thinner layers of high integrity plastic typically have greater measured breakdown strengths, this could result in an overestimate of breakdown voltage where test results are applied to greater layer thicknesses. An overestimate of breakdown voltage leads to an overestimate of PBD hazards (2-6.5).

Breakdown strength is lowered by pores, pinholes, or other spaces such as between the fibers in fabrics and composites. The corresponding reduction of breakdown voltage can in many cases prevent the occurrence of PBDs (2-6.5). In any case, breakdown strength and breakdown voltage should be determined by test rather than estimated.

2-6.2.5. Dust Ignition via Transitional Brushes

Transitional brush discharges have little significance in gas ignition since it can be assumed that ignition requirements are exceeded at the brush discharge stage. However, the situation is different for dusts since it is often assumed that PBDs have an effective energy of the order 1000 mJ and can therefore ignite almost any ignitable dust. PBD ignition tests of anthraquinone dust having a MIE of 2–10 mJ were carried out using a circular, 17.8-cm (7 in.) diameter, 150-μm-thick PVC layer. The surface charge density of ~5000 $\mu C/m^2$ corresponded to about 125 μC of stored charge [181]. It was found that some tests did not give ignitions and subsequent investigation showed that the PVC layer had failed to completely discharge during these tests. This suggests that poor dispersion of the dust was not entirely responsible for the lack of ignitions and that transitional brush discharges having a large range of effective energies were probably involved.

While all available evidence indicates that brush discharges from isolated nonconductors are incapable of igniting combustible dusts in air, Figure 2-6.2.3 shows that nonconductive layers represent a different case. Any dust whose MIE is exceeded by a PBD will also be susceptible to ignition by a transitional brush discharge occurring at some smaller layer voltage. If a fully developed PBD is not generated, the effective energy may be anywhere above the 5–10 mJ maximum for brush discharges from isolated nonconductors. As a consequence, only easily ignitable dusts might be at risk of ignition. Since this behavior is difficult to quantify it is appropriate in some cases to consider transitional brushes and PBDs as a single phenomenon as shown in Figure 2-6.5. In any case, it is necessary for large surface charge densities to be accumulated such as via an extended period of powder impact on the layer.

2-6.3. *Bulking Brush Discharge*

See 6-3.1.1 and 6-4.2. This is a large discharge resulting when a dispersed, charged powder "bulks" when settling in a container, causing a very large increase in its volumetric charge density. Infrequent surface flashes up to several feet long are observed in large containers being filled with powder having a resistivity above 10^{10} Ω-m, both during, and occasionally for a short time after, the transfer of powder. The discharges originate at the container wall and propagate across the bed surface. For axial powder feed the discharges appear between the edge of the powder cone and the surrounding walls (Plate 3) while for off-axis powder feed the discharges appear on the side opposite the powder cone [121]. An accompanying crackling sound has been heard from the top of a silo over the noise of the powder transfer. Transferred charge channeled to the wall may attain 25–50 μC [50,160] and are accompanied by large peak currents (6-4.1.1). Bulking brushes have an apparent maximum effective energy of 10–20 mJ (with respect to dust ignition) and are believed to be responsible for dust explosions in grounded silos. This is inferred from analyses of silo explosions where ungrounded objects and other ignition sources could be confidently eliminated. Hybrid mixtures (6-1.3) and dusts whose ignition energies are less than *Lycopodium* (see "Glossary") should be considered at risk from these discharges.

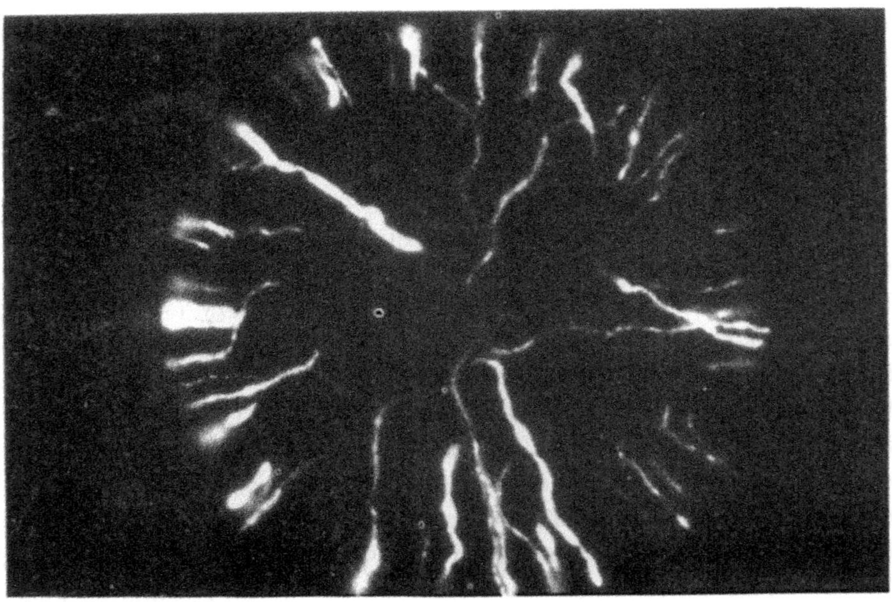

PLATE 3. Multiple bulking brush discharges on powder bed (time exposure). (From Bartknecht in "Industrial Dust Explosions," ASTM STP 958, p. 175)

2-6.3.1. *Effect of Particle Size*

It was initially thought that these discharges only occur during bulking of coarse powders with diameter 1–10 mm, such as pellets [25]. Later work demonstrated that this is not the case although the phenomenon is more frequently observed with coarse powders (6-3.1.1 and 6-4.2.1).

2-6.3.2. *Effect of Container Size*

Tests using vertical cylindrical silos have shown that the charge transferred by these discharges increases with silo diameter [161]. The maximum observed charge transfer Q varied with silo diameter D according to

$$Q = 2.3D^{2.8}\ \mu\text{C} \tag{2-6.3.2}$$

A wide distribution of charge transfers was observed with the maximum values roughly twice the average values. From Eq. (2-6.3.2) the maximum predicted charge transfer is about 40 μC in a 3-m-diameter silo and about 2 μC in a 1-m-diameter bin. The effective energy of discharges having comparable power densities is roughly proportional to charge transfer (2-5.1.1). If the maximum effective energy with respect to dust cloud ignition is assumed to be about 20 mJ (2-6.3) it follows that in small containers less than 1 m diameter the effective energy should be only about 1 mJ. This implies that the ignition hazard in small containers should be similar to that of a brush discharge, hence dusts in air should not be at risk of ignition. This result is consistent with the lack of reported ignitions when loading FIBCs (6-7).

The minimum size of container for bulking brush discharges has not been established, but is probably about 1 m^3. Their occurrence in such small containers might require coarse particles such as pellets and therefore represent a hazard only where flammable gas or hybrid mixtures are present. Assuming that the effective energy of the discharges is at the low end of the proposed 1–20 mJ range, the only significant feature vis-à-vis brush discharges is that no field-intensifying electrode is required to initiate the discharge.

2-6.3.3. *Occurrence Inside Powder Heap*

Bulking brush discharges occur not just on the powder surface but also deep within powder heaps, as demonstrated by large-scale tests [161]. This is one reason the author prefers the term "bulking brush" over "cone discharge" as used in some European texts; the other reason is that the discharges tend not to occur on the powder cone itself (2-6.3). Bulking brush discharges within the powder heap are usually nonhazardous because powders cannot combust in bulk. However, some unstable powders may be at risk since propagating decomposition is facilitated in the bulk state; this potential hazard with respect to bulking brush discharges has not previously been recognized (6-1.4).

2-6.3.4. Effective Energy Tests Employing Spark Ignition Chamber

Since the prevention of bulking brush discharges in large containers is typically impractical, it is extremely important to establish the proper magnitude of the effective energy. Some companies have been unduly alarmed by tests [161] suggesting that bulking brush discharges have effective energies up to 1000 mJ or more, implying that large silos containing ignitable dust suspensions may be accidents waiting to happen. The tests were made by channeling the charge collected from bulking brush discharges through a small spark gap in a Hartmann tube, a technique that has also been used to imply that brush discharges can ignite sensitive dusts [34]. As discussed in 2-5.1.1 and 2-6.2.1, this technique is unrealistic since it does not address the power densities of brush or bulking brush discharges. The tests demonstrated only that sparks transferring charge in the range 0.2–4.0 μC can ignite dusts having MIEs in the range $1 <$ MIE < 100 mJ under the specific test conditions used. The results appear to have been additionally biased by the assumption of the high MIE value using the "step method" (3-5.4.2), so that *Lycopodium*'s MIE of 30–100 mJ was interpreted as 100 mJ. Numerous other MIE evaluations give *Lycopodium*'s MIE as approximately 20 mJ.

2-6.3.5. Ignition Probability Considerations

Assuming that the effective energy of bulking brush discharges with respect to dust ignition is a continuum from about 1 mJ up to maybe 10–20 mJ, the infrequent occurrence of silo explosions even where easily ignitable dust is present suggests that most discharges have effective energies at the low end of this range (6–5). The probability of ignition is reduced by the need for an optimum mixture of fine dust in air to encounter the hottest part of the discharge. This occurs at the silo wall (2-5.1.1). After ignition takes place locally, the flame must propagate into the surrounding dust suspension without being quenched. It is additionally possible that the incendivity of these discharges depends on the polarity of the powder bed. Corona and brush discharges occur at a smaller frequency and carry greater effective energy when the charged nonconductor has negative polarity. Recent studies using pellets have concluded either that net polarity has a small effect on the occurrence of bulking brushes [50] or that net positive charging gives a greater frequency of bulking brushes [161]. The interpretation of polarity effects is complicated by the occurrence of bipolar charging. It is usually found that fine particles charge with opposite polarity to the coarse powder predominating in the powder heap. The net polarity is likely to be determined by the sign of charge carried by the predominant coarse particles. A net "positive" charge on a polydispersed powder stream does not rule out local accumulation of negative charge in the receiving container. To create a positive net charging current the positive charge need only be incrementally greater than the negative charge. As the predominant coarse powder settles, the suspended powder might contain a high net negative charge. Similarly, parts of the bed away from the cone might contain a high net negative charge. Tests in a 100-m^3 silo showed that when a powder stream with net positive charge entered near the wall, bulking brushes formed exclusively at the opposite wall [121]. A possible interpretation is that as coarse powder rapidly bulked in the off center powder heap, positive charge was released in the form of corona and brush discharges. The high rate of ionization created by these discharges suppressed formation of bulking brush discharges on the heap. It is unclear whether the appearance of bulking brushes at the opposite wall was simply due to the absence of a local ionized layer or whether the settling of powder having opposite polarity played a role.

It might be irrelevant that bulking brushes apparently occur more frequently with positively charged powder. Since coarse powders such as pellets cannot be ignited, what matters is the occurrence of bulking brushes in the presence of fine particles. It was shown that addition of fine powder to a silo containing a positively charged polyethylene pellet bed suppressed bulking brushes because the fine particles charged negatively and neutralized the bed [160]. As discussed in 6-3.1.1, if fines tend to charge negatively, ignition via bulking brushes might require predominantly negative charging to avoid such neutralization. An unrelated polarity effect that could influence ignition probability is maintenance of a fines suspension. Negative fines might quickly be precipitated on a positively charged bed and hence removed from suspension. Since ignition must occur at the level of the bed surface, fines removal via precipitation should decrease ignition probability. Again this suggests that if the fines are charged negatively, ignition via bulking brushes might require predominantly negative charging. Bulking brushes are discussed further in 6-3.1.1 and 6-4.2.

2-6.4. Spark Discharge

This is a transient discrete electric discharge which takes place between two conductors which are at different potentials, bridging the gap in the form of a single ionization channel (Plate 4). Based on light emission measurements of sparks with symmetrical electrode geometry, the energy is dissipated approximately uniformly along the channel. This is in contrast with asym-

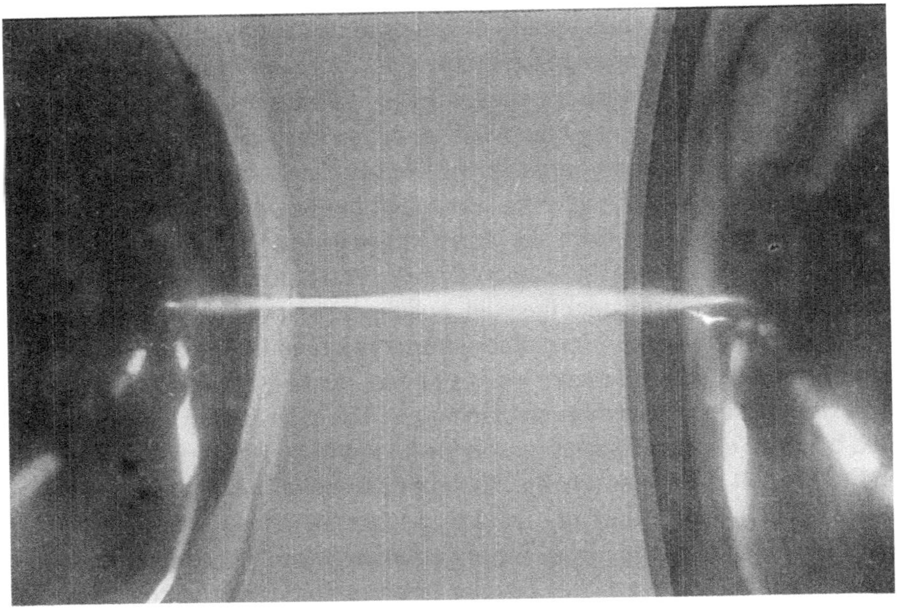

PLATE 4. Spark discharge between spherical electrodes.

metric sparks which normally exhibit a hotspot close to the cathode [155]. Since sparks are used to determine ignition energies, the effective energy is normally assumed equal to the stored circuit energy, equal to one-half the product of circuit capacitance and sparking voltage squared (C-1.2).

In reality only a small fraction of the energy stored in a spark circuit contributes to ignition. In addition to energy losses caused by nonidealities in the electrical circuit, only part of the energy released in the spark gap is usefully transferred. Energy is lost in the form of shock waves, dissipated as heat at the electrodes, and (in the case of long sparks) may be released outside the thermalization volume in which ignition occurs. The principal variables are spark gap length, spark duration, and electrode geometry, all of which possess optimum values for igniting a given flammable mixture. Additional variables include turbulence and electrode material [42].

Owing to loss of heat and radicals to the surrounding solid surfaces, flame propagation cannot occur in gaps smaller than the "quenching distance," which is a characteristic of the flammable mixture directly related to its MIE. For parallel plane geometry as described below, the relationship for hydrocarbons is given by MIE = $0.06d^2$, where the quenching distance d is given in millimeters and the MIE in milliJoules [224]. It is sometimes incorrectly stated in the literature that ignition cannot occur when spark electrodes are closer together than the quenching distance. However, this is only true where the quenching distance is properly defined by parallel plane geometry, such as where both electrodes are fitted with flanges (glass etc.) around their tips. In this case, ignition does not occur until the gap is slightly greater than the quenching distance. This gap length best approximates to a point source and yields the lowest MIE values. As the gap is increased beyond the quenching distance the spark becomes better described as a line source, which is less effective in producing ignition [208]. At larger gap lengths the MIE therefore increases from its lowest value. In the absence of parallel plane geometry the concept of quenching distance does not directly apply. If either electrode is unflanged and approximates to a point, ignition can occur in gaps much smaller than the quenching distance. Although MIE is increased, a gap smaller than the quenching distance does not prevent ignition in the general case.

2-6.4.1. Spark Breakdown

Paschen's law states that the sparking voltage is a function of gas density and gap length alone. The law is usually followed for gas pressures up to a few atmospheres with small dependence on temperature. The sparking voltage in a uniform electric field is known as breakdown voltage or breakdown field. It should be noted that corona and brush discharges may occur at voltages considerably less than the sparking voltage. At atmospheric pressure, the breakdown field of air varies continuously from 45 kV/cm at a gap of 1 mm to 26 kV/cm at a gap of 10 cm. An "average" value of 30 kV/cm, corresponding to a gap of 2 cm, is often used as the approximate breakdown field of air. Atmospheric air, dry air, nitrogen, carbon monoxide, acetone vapor, and methane all have roughly the same breakdown field. Oxygen and carbon dioxide have 85–90% of the value while hydrogen has about half the value and rare gases such as helium about one-tenth of the value. For organic compounds in an homologous series, such as methane through octane, the breakdown field usually increases with increased molecular weight with little effect due to branching. The presence of double and triple bonds also increases breakdown field relative to aliphatic hydrocarbons. When hydrogen in a molecule is replaced by a halogen the breakdown voltage is usually increased; an example is carbon tetrachloride, which has a breakdown field about 170% that of methane or air. Sulfur hexafluoride has been used industrially to prevent charge loss via static discharge, having about 240% the breakdown field of air. The effect of halogens is due to their high electronegativity, facilitating electron attachment to form relatively massive negative ions. Oxygen and water are also electronegative, but less so than halogens; their effect on breakdown field on addition to dry air is measurable but relatively small. For most practical situations involving flammable vapor concentrations, the breakdown field is essentially determined by the properties of air. Exceptions may occur for vapors or gases having unusually wide flammability limits, such as hydrogen.

The significance of this discussion is that the appearance of a spark may, in addition to charge accumulation rate, involve changes in gas density, gas composition and electrode geometry. Release of flammable gas into an electrically stressed region may initiate spark discharge. An extreme example is lightning striking an atmospheric vent stack. If lightning does not strike the stack, ignition might still occur via brush discharge or via an upward streamer that fails to connect with the downcoming stepped leader from the cloud (see 5-9.4). Lightning strikes are notoriously capricious and often do not strike "obvious" targets. Were this not the case a giraffe would truly be sticking its neck out during a thunderstorm on the savanna.

2-6.5. *Propagating Brush Discharge (PBD)*

This is a very energetic discharge produced by breakdown of an electrical double layer (capacitor) and is a special case of the phenomenon described in 2-6.2.2. The principal difference is that under certain conditions the electric field parallel to the non conductive surface becomes sufficiently great for lateral flashover to occur, enabling much of the surface charge to be collected in the discharge (A-2-6.5). The maximum effective energy of a PBD is probably about 1000 mJ (2-6.5.3), although "transitional" cases can have much smaller energies depending on the charged surface area and extent of discharge (2-6.2.3). From an ignition perspective a PBD might be distinguished from a brush discharge by the occurrence of a minimum charge transfer of 10 μC, roughly an order of magnitude larger than the maximum charge transferred by a positive brush discharge. Fully developed PBDs on extensive surfaces with large breakdown voltages typically transfer at least 100 μC of charge. They are accompanied by a bright flash and a loud, sharp report comparable to a rifle shot.

PBDs were first observed in 1777 by G. Chr. Lichtenberg and are sometimes known as "Lichtenberg discharges." A review of Lichtenberg's electrophorus work and the "Lichtenberg" powder patterns obtained is given in [187]. These powder patterns are produced by scattering dust over a surface upon which a PBD has recently occurred and give a remarkably clear depiction of the intricate discharge channels. The patterns are found to depend on the polarity and surface charge density. Oak Ridge National Laboratory produced a discharge resembling a PBD after irradiating Pb–Ce glass with the 10 μA electron beam from a 1.5 meV van de Graaf accelerator and then striking the edge of the glass with a sharp point; the beam penetration was about 1 mm into the 1-cm-thick glass. The discharge propagated from the stressed point parallel to the surface within the 1 mm layer [188].

To understand the mechanism of PBD production, consider powder impact on an insulating layer with grounded metal underneath. Charge accumulation on the insulating layer is balanced by an equal and opposite countercharge induced into the underlying metal surface. Since this creates a capacitor with zero net charge, the external field is due only to the difference in position of the two charged surfaces relative to an object above the insulating layer. The external field increases with the thickness of the insulating layer and for thicknesses above about 8 mm air breakdown occurs before the critical surface charge density of 2.5×10^{-4} C/m^2 can be attained [36]. Charge otherwise continues to accumulate on the insulating layer until the breakdown voltage of the layer is attained. When the insulating layer breaks down directly under the internal electric field, layer puncture is accompanied by massive lateral discharge to the puncture point owing to very large surface potential gradients. Alternatively, PBDs may be initiated by mechanical stress or by approach of a grounded electrode to the insulating layer. In the latter case, gas breakdown occurs on the approach of an electrode provided the net electric field at the electrode is sufficient. This in turn triggers the PBD (Plate 5). The phenomenon is affected by the rate of charging, the duration of charging, the breakdown strength of the layer and the layer thickness.

PBDs were not observed for layers less than 23 μm thick and the PBD film potential threshold varied from 4–5 kV at 23 μm to 11 kV at 175 μm [228]. A later study, possibly using a less valid method for measuring film potential, reported smaller threshold values [29]. The later study concluded that if film breakdown voltage is less than about 4 kV at 20 μm and about 8 kV at 200 μm, PBDs cannot be produced. This is shown schematically in Figure 2-6.5.

As discussed in 2-6.2.3, the figure has been conservatively drawn and the indicated PBD region includes both brush and transitional brush discharges. The voltage thresholds for fully developed PBDs are therefore higher than indicated. As discussed in 2-6.2.4 and Note 2 of 2-6.5.1, the breakdown voltage line corresponds to an envelope of maximum values for various films rather than specifically to polycarbonate film. The values correspond to breakdown strengths a factor 2–10 greater than those given for plastic films in Appendix B.

2-6.5.1. *Principal Criteria for PBDs*

The principal criteria under which a PBD can be produced on a nonconductive layer are

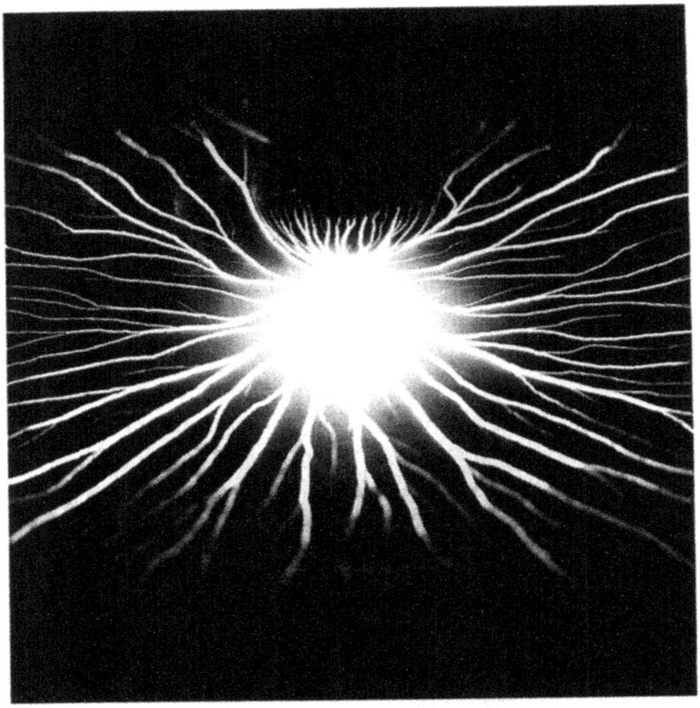

PLATE 5. Propagating brush discharge (PBD) on charged layer initiated by grounded electrode.

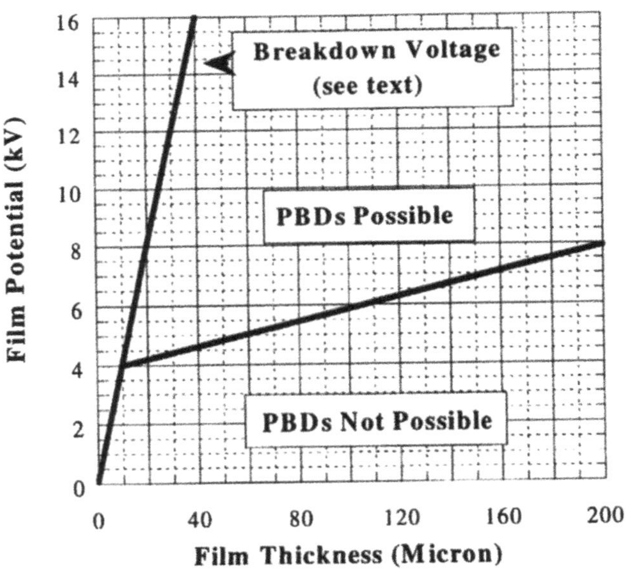

Figure 2-6.5 . Schematic conditions for PBDs (adapted from [29] see text)

1. The layer thickness must be less than about 8 mm or air breakdown will occur above the charged layer before the critical surface charge density of 2.5×10^{-4} C/m^2 can be attained.
2. The breakdown strength of the charged layer must be adequate to attain the critical charge density, implying a certain minimum layer thickness as shown schematically in Figure 2-6.5.
3. Sufficient charge must be available to supply the required critical surface charge density. Calculations show that 50-pound plastic bags and plastic lined 55-gallon drums are too small to represent a credible risk.

Note 1: As discussed in 2-6.2.3, Figure 2-6.5 has been conservatively drawn to accommodate not only PBDs, defined by a minimum charge transfer of 10 μC, but also brush discharges having charge transfers less than 1 μC. Charge transfer and discharge energy both increase with voltage across the layer, and this is limited by the layer breakdown voltage. Figure 2-6.5 shows that PBDs can in all cases be prevented by keeping the layer breakdown voltage less than 4 kV. However, greater breakdown voltages are allowed for thicker layers and as noted in 2-6.5, an earlier study [228] reported PBD film potential thresholds that are higher than indicated in Figure 2-6.5.. Apart from simple rubbing applications (2-6.2.2) the selection of a "safe" layer thickness has little application for gases since brush discharges are themselves an ignition hazard. However there is valid application for dusts (2-6.2.3 and 6-4.3).

Note 2: As discussed in 2-6.2.4 the published breakdown strength of most plastics is less than the 400 V/μm indicated in Figure 2-6.5. Values for capacitor-grade polyethylene and PTFE films are 39 and 59 V/μm respectively [183]. The breakdown voltage line shown in Figure 2-6.5 was determined by enveloping the maximum breakdown voltages measured for various thicknesses of plastic film or combinations of films up to about 50 μm (2 mils) thick. It was concluded in [29] that breakdown strengths differ only slightly between different plastic films, despite the more than 1 order of magnitude range of values cited in [183]. As discussed in 2-6.2.4 the breakdown voltage of thick plastic layers can be much less than suggested by Figure 2-6.5.

2-6.5.2. PBDs from Isolated Nonconductors

PBDs may occur on plastic surfaces with no metal substrate, for example the wall of a plastic pipe conveying charged material. In this case the double layer forms between the inner charged wall of the pipe and a countercharge which accumulates on the outer wall via conduction or via corona discharge. In the latter case both layers of charge reside on nonconductive surfaces and a PBD on one side of a plastic surface should be accompanied by an equal and opposite PBD as the countercharge discharges to the same puncture point. Following an acrylic powder explosion in a railroad bulk container, meter long PBDs were directly observed in a short section of polyethylene pipe running to the container in an otherwise steel line; it was found that the plastic pipe outer wall had been rendered conductive by moisture and was effectively grounded at several locations. The charged inner wall and grounded outer wall formed a capacitor [13]. A large PBD was photographed on a section of 2-in.-diameter plastic pipe conveying charged diesel oil, where the outer wall was dry but contaminated with dirt and oil residues [8]. The photograph shows a roughly meter-long discharge initiating at a puncure point through the wall. Diesel oil was evaporated and ignited external to the pipe (Plate 6). Following an explosion, a PBD was concluded to have occurred on a section of plastic pipe in a chocolate crumb silo whose outer wall had been wrapped with grounded, conductive tape "to help dissipate static" [14]. Instead, the grounded tape actually contributed to the formation of a capacitor on the plastic pipe. As a final example, a PBD was reported in a 5-mm-thick walled polyethylene tote bin filled with charged, 1-mm polypropylene granules. The bin's metal support frame supplied countercharge to the outside plastic wall via corona. In this last case, a worker reaching into the bin could short out the capacitor that had formed and receive a severe shock [15].

2-6.5.3. *Stored PBD Energy*

If it is assumed that PBDs transfer a minimum of 10 μC of charge across a layer with breakdown voltage 10 kV, the minimum stored energy is of the order 100 mJ. The maximum stored energy is limited by the surface area and breakdown voltage V_b of the insulator. For large surfaces an additional limiting factor is the maximum flashover distance; if this is assumed to be about 0.5 m (A-2-6.5), the maximum dimension to consider for layer diameter or pipe length is 1 m. From Appendix C-1.2 the stored energy W is

$$W = CV_b^{\,2}/2 = QV_b/2$$

where Q is the product of surface area and surface charge density, and C is capacitance given by

$$\text{For a layer} \quad C = A\,\varepsilon_r \varepsilon_0/d$$

where A = area and d = layer thickness.

$$\text{For a pipe} \quad C = 2\pi l \varepsilon_r \varepsilon_0/\ln(r_2/r_1)$$

where l = length and r_1, r_2 = inner and outer radii.

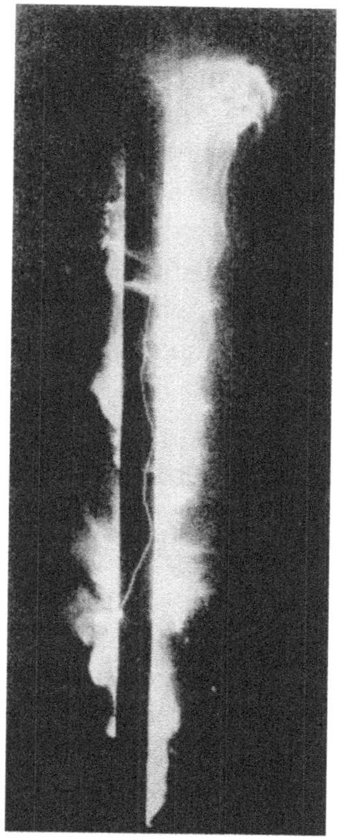

PLATE 6. Approximately 1-m-long spontaneous PBD on plastic pipe.

Since V_b for 1 mm of thickness may be of the order 20 kV (Appendix B) it follows that stored energies can easily achieve 1–10 J. It can be assumed that ignition occurs close to a puncture point through the layer, into which the energy is channeled. Since a large fraction of the stored energy is released remote from the puncture point(s) and cannot contribute to ignition, the maximum effective energy is probably about 1000 mJ.

2-6.6. *Surface Streamer*

This is the author's classification for a large surface discharge observed on charged liquids, particularly during road tanker filling, appearing as a flash up to about 0.6 m long and accompanied by a crackling sound. It is not described in current electrostatic texts but was commonly discussed in US publications of the early 1960s, being referred to as the "go-devil" [61] or as "severe sparking" [148]. Its effective energy is unknown, but should be greater than that of the brush discharge; its similarity to the bulking brush discharge suggests a similar effective energy of at least 10 mJ. High charge densities from an upstream filter with insufficient residence time (5-3.5) plus a liquid with very low conductivity might produce this wall-to-liquid surface discharge. An initiating mechanism might be transient convection of abnormally high charge densities to the liquid surface, possibly influenced by charged patches of froth. It is assumed that the large propagation distance of surface streamers and bulking brushes is due to the type of step-by-step mechanism described in A-2-6.5, which allows discharge lengths much greater than would be possible for sparks in air gaps at the same voltages.

2-7. Personnel Spark and Shock Hazards

These are normally due to charging of people and subsequent discharge in the form of a spark to a conductor. The energy dissipated (W, Joules) approximates to the discharge of a capacitor (C, Farad) having the same capacitance as the person and charged to the same voltage (V, Volts), or $W = 0.5CV^2$. Common "doorknob" shocks caused by walking across carpet or after getting out of upholstered chairs are of this type. Shocks can also result from static discharges from charged items of equipment. In the case of sparks, the energy can be calculated directly if the capacitance and voltage of the charged conductor are known. Other types of discharge from charged bodies, with the exception of the propagating brush, will not give significant shocks and in many cases a discharge capable of igniting gas mixtures may be imperceptible. The propagating brush (2-6.5) may be initiated by touching a nonconductor charged in the manner of a capacitor. PBDs commonly have stored energies of about 1 J (2-6.5.3) and in extreme cases could give hazardous shocks approaching the lethal threshold of about 10 J (2-7.3).

2-7.1. *Body Capacitance and Resistance*

The capacitance of a person is not constant but varies by about a factor of four depending on factors such as the type of shoe, attitude with respect to ground and proximity to neighboring conductors. Human body capacitance is unlikely to be less than 90 pF, corresponding to a person of average height standing on the tip of one rubber boot with 18-mm sole (simulating walking with the smallest contact area with the floor), although an individual who is completely isolated by jumping off the floor may achieve 55 pF. A typical value for a person in standing position is 120 pF. In general, the range of body capacitance can be taken as 100–400 pF with 200 pF frequently used as an average value. Higher values correspond to unusually high shoe capacitance with respect to ground (such as thin, flat, nonconductive soles on a conductive floor). Body capacitance in this case can be calculated using the equation in C-1.1 for two parallel capacitor plates ($C = A\varepsilon_r\varepsilon_0/d$, total foot area A, sole thickness d). Total body capacitance may be further increased by proximity to large vertical conductors such as steel tanks. The body resistance of an average person, as measured at 30 V between one finger and two bare feet in contact with a metal plate, varies from about $4 \times 10^4\ \Omega$ (from ball of finger) to $2 \times 10^5\ \Omega$ (from tip of finger), where the exact value varies with skin thickness and dryness [53].

2-7.2. Voltage (V) and Energy (W) Attained

For the purposes of discussion it is assumed that a person has a capacitance of 200 pF. People are most unlikely to achieve voltages exceeding 50 kV, which might correspond to standing in a pile of freshly dumped, highly charged thermoplastic resin (W = 250 mJ). When sliding over carpet in dry offices, voltages up to about 30 kV have been observed (W = 90 mJ). At low values of relative humidity, walking across carpet or getting out of an upholstered chair often results in body voltages up to about 10 kV (W = 10 mJ). Such voltages may be attained even when walking over painted stone floors provided shoe resistivity is high. Since the level of perception is about 1 mJ it can be appreciated that human body voltages exceeding 3000 volts are very commonly attained. Appendix B shows that 1 mJ exceeds the MIE of many gas mixtures in air. Sparks from people have a smaller effective energy for ignition than those between small metal electrodes. This is due to a larger capacitance and effective electrode radius (Figure 3-5.4), energy lost in skin resistance, plus a tendency to continue releasing energy in a series of smaller sparks following an initial large spark [38]. Even though the effective energy may only be 20% or less of an optimized metal–metal spark having the same stored energy, ignition of gas mixtures by charged people may easily occur unless precautions are taken (4-3). Dust ignition is possible although less likely owing to the higher MIEs of dust clouds plus the smaller probability of the spark encountering an easily ignitable mixture. Considering the depressed effective energy of sparks from people and the fact that shocks more severe than that from a spark plug (2-7.3) are extremely unusual, flammable mixtures with a MIE above the range 30–50 mJ are probably not subject to this ignition hazard. In flammable gas atmospheres, sparks from people may be most hazardous where they cause a borderline response and where the situation may therefore continue uncorrected.

2-7.3. Human Shock Response

This depends on the energy and source capacitance of the spark plus personal characteristics including skin resistance, sensitivity and general health. Response to mild shocks is related to the power density passing through the skin. Hence if one is carrying a metal key a larger energy can be dissipated without shock when the key is used to discharge the body to ground. In practical situations involving metal tools this can result in production of sparks of at least several millijoules without any perception of a shock, and the sparks may not be either audible or visible. The following are typical response thresholds for capacitance sparks to the bare hand: 0.5–2 mJ (perceptible), 1-10 mJ (various levels of discomfort), 15–25 mJ (unpleasant shock), 250 mJ (severe shock), 1–10 J (possible unconsciousness), >10 J (possible cardiac arrest). As a reference, a typical automobile spark plug dissipates energy in the 25-35 mJ range, with extreme values (integral of plug VI over time) of 10–50 mJ [48]. Although a shock may be physiologically nonhazardous, the involuntary effects may be extremely hazardous in causing falls or other indirect injuries. The latter includes third party injury from a dropped item, spillage of open or breakable containers and impact laceration from machinery or sharp objects.

3

EVALUATING THE HAZARD OF STATIC ELECTRICITY

3-1. General

It is recommended that flammability always be assessed first, since it is inherently safer to avoid flammable atmospheres than to avoid sources of ignition such as static electricity. If a flammable atmosphere cannot be avoided at all times, the system should be designed to minimize both the probability and consequences of ignition. In this chapter it is assumed that static electricity is the only source of ignition; however, in practical situations all sources of ignition such as those described in [157] should be evaluated.

3-2. Hazard Identification Methods

This process is carried out by analyzing the following

- Can a flammable atmosphere be created?
- Can static electricity be generated rapidly enough to accumulate?
- What type of static discharge might occur?
- Could this discharge carry an effective energy sufficient to ignite the flammable atmosphere?

The potential to form a flammable atmosphere is addressed using methods discussed in Chapters 5 and 6.

3-2.1. Decision Trees

It is often helpful to use a decision tree for routine hazard analyses. This structured approach ensures that a certain minimum number of analytical steps must be followed, depending on the complexity of the tree. One weakness with the approach is that a series of yes/no decisions must be made in order to progress. Owing to the complexity of the subject, decision trees attempt to use conservative decision criteria. It is far easier to determine the nonexistence of a specified hazard than the existence of one. A second weakness is where the analysis replaces on-site inspection. This can be mitigated to some degree by the presence of a knowledgeable operator or process engineer, but ideally the decision tree should be used to identify potential hazards that are later investigated on site. An explosion may result from a minor process change that affects either the accumulation of static or the formation of a flammable atmosphere. This may, for example, involve improper replacement of a single component or the unrecognized impact of recent compliance with clean air regulations. Useful adjuncts to a routine hazard analysis are (1) collection of incident and near-miss cases both for the process and similar processes and (2) to assume that a serious fire or explosion has taken place. By encouraging lateral thinking or "thinking outside the box," items not included in the decision tree might be uncovered.

Decision trees are not used in this book since they are most useful when targeted to a specific process; attempts to generate comprehensive matrices rapidly lead to extremely complex schemes. Instead, the book should be used to help generate suitable matrices or to supplement the decision-making steps in published matrices such as [199]. Many of the decision steps, such as the conditions under which discharges of some specified effective energy may occur, are not properly understood and continue to be controversial.

3-3. Charge Accumulation

3-3.1. Conductive Objects

Conductive objects may accumulate charge via conduction, induction, ion collection or contact-separation in relation to a nonconductor. In general, charge accumulation on conductive objects can be completely prevented by bonding and grounding. The only decisions to be made are how small the resistance to ground must be (4-1.3) and how small a conductive object needs to be before it can be neglected. The latter depends on the results of the hazard analysis (3-2) and in some cases only objects commensurate with nuts and bolts (about 3 pF) can be neglected.

3-3.2. *Nonconductive Objects*

Accumulation of charge is usually the result of contact and separation of surfaces. In the case of powders both signs of charge are often produced simultaneously during flow and the net charge accumulated may depend on later classification of the powder in different parts of a handling system, as discussed in Chapter 6. Nonconductors can accumulate charge by ion collection such as via DC corona discharge. Significant induction charging does not occur. Charge accumulation on nonconductive objects cannot be prevented by bonding and grounding. In some cases the volume or surface conductivity of the object can be increased and in other cases it might be possible to prevent accumulation by charge neutralization techniques.

3-4. Energy Estimates

The energy stored on a conductor can readily be calculated from the object's capacitance and voltage (C-1.2). The capacitance may be directly measured using a capacitance bridge or may be estimated from the object's size and shape. A-4-1.3 gives some capacitances of common objects.

Actual capacitance values can change according to the proximity of other conductors. Consider an ungrounded sample thief with constant charge Q being removed from a tank. As it nears the grounded wall its capacitance increases. If its capacitance were to double its voltage would be halved ($V = Q/C$) and the stored energy W available as a spark would also be halved ($W = QV/2$). This effect can be neglected for a first approximation.

At fixed capacitance, energy increases with the square of the voltage ($W=0.5CV^2$) and voltage is therefore the dominant factor determining stored energy. In some cases the maximum voltage can be estimated from the geometry. For example, if the minimum distance to a grounded point is known this distance will determine the maximum voltage before spark breakdown occurs; in some cases such as accident investigation it might be necessary to accurately simulate the geometry to determine this voltage. Where the conductor is in the form of a container such as a bucket, the energy may be found from the container capacitance and charge inside the container (C-1.2). The charge may be estimated by Faraday pail experiments (3-5.1.2) or using estimates of liquid charging current (5-3.1.1). The estimated spark energy is then compared with the minimum spark ignition energy for the flammable material (3-5.4). Where static discharges other than sparks might be produced the "effective energy" of the discharge must be considered instead (2-5.1).

3-4.1. Charge Sharing

When a contact instrument is used to measure the charge or potential of a conductive object, charge is "shared" with the measuring device. For example a voltmeter and cable must draw charge from the object in order to attain the same potential. If C_0 is the capacitance of the object and C_M the capacitance of the measuring device plus cable, the original charge is distributed between capacitances $(C_0 + C_M)$. The original voltage V_0 is reduced to the measured voltage V_M. From the relationship $Q = CV$, the correction for voltage is

$$V_0 = V_M (1 + C_M/C_0) \qquad (3\text{-}4.1)$$

The correction can be large for objects having capacitances less than about 50 pF since C_M is typically in the range 10 to 20 pF (3-5.3.2 and 3-5.3.3). Neglect of charge sharing can result in a large underestimation of stored energy. This follows from the relationship $W = 0.5CV^2$ given in C-1.2. The calculated stored energy W_M based on measured V_M is much less than the original stored energy W_0 based on V_0

$$W_M / W_0 = \{1 / (1 + C_M/C_0)\}^2 \qquad (3\text{-}4.2)$$

3-5. Instrumentation

Of the instruments described only some are suitable for use outside the laboratory. Where any instrument is carried into a flammable environment it should either be certified as intrinsically safe for exposure to the flammable atmosphere or isolated from the atmosphere such as by keeping it within a purged enclosure. Any probe connected to the instrument must be separately considered as a possible ignition source. Electrometers are described in [153]. A more general review of electrostatic instruments is given in [136].

3-5.1. Charge

Total charge, volumetric charge density, and charge-to-mass ratio (specific charge) are useful parameters in assessing electrostatic hazards. Charge on conductors can be directly measured using a device such as an electrometer; however it is important to account for the charge "shared" with the measuring device's capacitance (3-4.1). When measuring the charge on nonconductors, indirect methods must normally be used. These either measure charge induced onto a conductor, such as in the Faraday pail method (3-5.1.2), or measure the electric field due to charge under well-defined geometrical conditions. An example of the latter is a field meter mounted flush with the inside wall of a grounded metal pipe containing charged, nonconductive liquid. In this case, the relationship given in C-1.4 for maximum electric field at the wall of a long cylinder can be used to find the charge density in the liquid.

The A. O. Smith charge density meter for liquids, which has been widely used by the petroleum industry, utilized a grounded, perforated metal sample cup extending into the pipe. The electric field at the end of the squat cylindrical sample cup was measured using a field-mill mounted on the pipe. This field was linearly related to the liquid charge density in the sample volume by Poisson's equation for a squat cylinder, or approximately, for a sphere (C-1.4). A drawback with this arrangement was additional charging of the liquid as it flowed through the perforations in the sample cup, preventing confident interpretation of small charge density readings. However, had the charge density meter been designed for flush wall mounting to avoid local charging by the cup, separate calibrations would have been needed for each pipe diameter and the device could only have been used in long pipe. Flush-mounting with in-situ calibration has been successfully used for experimental work [68].

3-5.1.1. *Electrometer Methods*

Most electrometers measure charge internally and are equipped with suitable input cables. Provided cables and connections are well maintained, charge leakage is extremely small and negligible during the time taken to make a test. If the quantity of charge exceeds the available range, an external capacitor can be used to store the charge and the electrometer can be used to measure the voltage across this capacitor ($Q = CV$). It is important to avoid stray capacitance using grounded screens where appropriate and to minimize current leakage through any external capacitor. The capacitance of the electrometer and cables must be accounted for.

3-5.1.2. *Faraday Pail*

This is used to measure charge, surface charge density, volumetric charge density or charge-to-mass ratio. It comprises an all-metal container, such as a 1 gallon can, isolated from ground with PTFE or other highly resistive plastic insulator. The pail is screened from external electric fields and changes in stray capacitance by isolating it within a larger, grounded metal container. Both the pail and the screen can be constructed either of metal sheet or metal mesh. If charged material is placed in the pail an equal charge is induced onto the walls of the inner container and this charge may be directly measured using an electrometer or other device. The charged material does not need to be conductive because induced charge immediately appears on the pail, unaffected by the rate at which charge dissipates from the sample. For example, a Faraday pail may be used to collect charged liquid or powder from a stream, yielding the total charge and subsequently (after weighing) the charge-to-mass ratio. To make a measurement the pail is connected to the charge measuring device and momentarily grounded to set zero. The charged sample is then introduced and the total charge recorded. The input capacitance of an electrometer is typically much larger than that of the pail plus cable, so that effectively all the charge is measured by the electrometer.

3-5.1.3. Charge Measurement via Filtration of Particles

A Faraday pail operating at negative pressure can be used to sample mist or dust at known flow rate and hence determine the volumetric charge density. A system for evaluating charge density in water mist during tank washing comprised an insulated metal chamber filled with filtration media, surrounded by a grounded coaxial metal screen. The charge density of the water mist was found by dividing the current appearing on the filter chamber by the flow rate. Current was measured by an electrometer [84]. When designing this type of device it is important to minimize particle impaction on any sample tube or external screen prior to entering the measuring chamber. Sampling tubes are normally not practical and the device must normally be placed at the sample location.

3-5.1.4. Charge Transferred in Static Discharges

In principle this can be easily found by integrating the current flowing to ground, typically by applying the voltage developed across a small resistor in the ground return line to the input of an oscilloscope or high speed digital recorder. Alternatively, the charge can be calculated from the voltage developed across a capacitor in the ground return line. Electrometers allow this to be done directly; the internal capacitance is sufficiently large for the electrometer input to remain close to ground potential. Image charge effects at the electrode can cause the charge transfer to be underestimated (C-2.3) but for small electrodes used for brush discharge measurements the error is typically of the order 10% and can often be neglected. This problem is discussed in [71,198]. Where the electrode is large, such as might be used for bulking brush discharge measurement inside silos, it is necessary to minimize image charge errors (C-2.4). Bulking brush measurements involve a number of experimental difficulties. For a series of large-scale tests a 0.5-m-high, ring-shaped foil electrode was used inside a silo to collect charge [160,161]. It was stated that corrections were made for image charge effects but the method used was not described [160]. In any case, large charge transfers were reported with less than factor-of-two differences between maximum and average values, suggesting that the foil electrode was appropriately sized for the silo involved. A simple, experimentally based correction method for image charge effects is discussed in [71].

3-5.1.5. *Surface Charge Density (Q_s or s)*

For small plane surfaces, such as test specimens charged on one side, this is most easily measured using a Faraday pail (3-5.1.2). The charged sample is dropped into the pail and the surface charge density calculated from the total charge and surface area of one side. For large or fixed surfaces the quantity is calculated from electric field measurements made using an electric field meter. If the meter head is equipped with a large grounded annulus such that the relative area of the sensing aperture is small, and the sensing head is brought close to the charged surface, the electric field is approximately uniform and is directly related to surface charge density (3-5.2.1). If a field meter sensing head with no grounded annulus is separated by a distance of several head diameters from a charged nonconductive surface, the electric field is intensified at the sensing aperture (C-2.5.3). The measured field may be several times greater than would be present under uniform field conditions.

It is important to recognize that even if the field meter head has a large annulus designed to allow uniform field measurements, the process of measurement may double the field that previously existed. As discussed in C-2.5.4 a charged, isolated, nonconductive surface exerts a normal electric field in both directions. The relative magnitude of these fields depends on the locations of all grounded conductive surfaces in the surroundings. If the field meter head is the only nearby grounded surface, the surface charge will couple with it and the field will all be exerted in its direction (C-2.5.5). Provided the surface is only charged on one side this makes no difference in the calculation of surface charge density, since the equation $E = (s/\varepsilon_0)$ already assumes a unidirectional field.

A particular problem is determination of surface charge density on a nonconductor which has charges of opposite polarity on either side (bipolar charge). The opposing fields created by these charges at the field meter sensor tend to cancel one another. Bipolar charging of webs is uncommon but the lack of a sensible field can defeat both field detection and the effectiveness of neutralizer devices. A fire incident is described in [16]. A method that can be used for surface charge density on webs having bipolar charge is to measure the electric field from the web as it passes over a grounded metal roller. The charge on the surface contacting the roller is coupled with the roller and produces no external electric field. The outside surface produces an electric field E which can be measured to yield the surface charge density s. For a web of dielectric constant ε_r, thickness t, and separation d from a parallel field meter head, the relationship is $E = st/(\varepsilon_0 \varepsilon_r d)$. For example, a field meter positioned 1 cm above a 0.1-mm-thick web having a dielectric constant of 2 and surface charge density of 1 μC/m^2 should in principle indicate an electric field of 564 V/m [136].

3-5.2. *Electric Field*

Appendix C-1.4 shows that electric field is related to charge and potential although the relationships are rarely simple. Complicating factors include nonuniform charge density and nonsimple boundary conditions. When an electric field is measured the measuring device, normally comprising a metal probe or sensor, reduces the local potential while intensifying the local field. Applications in large volumes such as tanks may require a correction factor for field intensification. However, when measuring the field due to plane arrays of charge, such as on plastic sheet, uniform field conditions may be approximated by using a separation between the sensor head and sheet that is small compared with the width of the sensor head. The latter can be equipped with a grounded metal annulus to help achieve parallel-plate geometry. A uniform field reading of 150 kV/m might indicate the potential for brush discharges in the presence of electrode which intensifies the unidirectional field by a factor of 20 (C-2.5.3). A uniform field reading above 300–400 kV/m is often indicative of a possible brush discharge hazard. Where field meters are used to measure surface charge density, for example on plastic film, it is difficult to avoid complications due to geometrical effects that, even with uniform charge density, may lead to errors of about a factor of 3. These problems are discussed in detail in [40].

Electric field measurements are often performed for qualitative or comparative applications only. It is quite common to interpret electric field readings as "surface potentials" and instruments are available for making "surface potential" readings on nonconductive surfaces such as belts and webs. As noted in 3-5.1.5 and 3-5.2.1 such measurements do not yield surface potentials and are usually not performed in a manner that would yield accurate surface charge densities. However, provided the instruments are used as recommended by the manufacturer and interpreted with respect to "surface potential" thresholds given in various recommended practices, the measurements are useful in determining whether a charging problem exists and in the assessment of static neutralizer systems. It is important to recognize the qualitative nature of many reported electric fields and "surface potentials."

3-5.2.1. *Field Mills and Vibrating Sensor Field Meters*

These are noncontact devices which respond to electric fields by measuring the charge induced by the field on an isolated metal sensor element. The voltage appearing on the sensor element is proportional to the product of sensor area and electric field. With a known capacitance between the sensor and ground the charge appears as a voltage across this capacitor which when input to a high impedance amplifier generates a signal proportional to the electric field. Owing to leakage resistance from the amplifier input this charge would normally rapidly leak away. However if the electric field is modulated by changing the position of a ground surface in front of the sensor, at a rate which is fast compared with the RC leakage time constant, an alternating signal is generated whose peak-to-peak amplitude accurately reflects the electric field strength and is independent of modulation frequency. The field mill modulates the signal using an external segmented chopper vane, or "mill". The polarity of an external electric field can be determined using a subsidiary chopping vane internal to the instrument which generates a reference signal with the same modulation frequency. The two signals can be used to assign polarity using a differential amplifier. The instrument is set up so that maximum signal corresponds to maximum sensor exposure to the field. Vibrating sensor field meters operate on a similar principle and various configurations exist. Usually these offer smaller sensor and hence probe dimensions than field mills. Field mill motors can be gas driven and special features of both types may allow operation in hazardous locations. Some types will operate in nonconductive liquids.

When any field meter is used in atmospheres containing charged dust, measures must be taken to prevent deposition of dust near the sensing aperture, particularly if the dust is nonconductive. Attraction and deposition of dust near the sensor gives a zero offset which can introduce an intolerable error. Air purging and frequent zero checks have been recommended [136]. Short of removing the meter head to a field-free region, a zero check might be accomplished using a grounded shutter [12]. A field meter used in a pneumatic powder pipeline employed inert gas to supply both a purge stream and a high velocity "air knife" effect across the sensor to prevent deposition of charged dust [12].

The applications of field meters are discussed in [136, 217]. Typical uses are

1. Electric field measurement at the boundary of a metal container filled with charged material. Examples include pipelines and storage vessels. The electric field can be used to calculate charge density (3-5.1). Field meters can also be lowered into containers such as silos to determine the local fields and polarities. Quantitative interpretation of the reading requires correction for field intensification and is sometimes accomplished using computer simulations.
2. Measurement of space potential in a large container using the field meter head as a potential probe (3-5.3.4).
3. Surface charge density measurement on nonconductors. If the field meter sensing head is mounted within a relatively large, grounded plane surface which is brought within several tens of millimeters from the charged surface, the surface charge density is related to electric field by $E = (s/\varepsilon_0)$. This is discussed further in 3-5.1.5 and C-2.5.5.
4. Surface potential measurement.

Provided the field meter head is equipped with a grounded annulus to produce a uniform electric field, the surface potential on a plane conductor is directly related to electric field and separation from the meter head. The relationship is $V = Ed$ where d is the separation between the two parallel conductive surfaces. The uniform field requirement is similar to the previous case (3). The field meter is usually calibrated by applying a known potential to a metal plate using a constant voltage power supply. Where uniform field conditions are not expected during use, the calibration can be specifically carried out for the fieldmeter head at a prescribed separation. The field meter can be calibrated directly in volts for a given separation or a calibration curve can be constructed for a range of separations. The target surface must be large compared with the separation to avoid having to make a further correction for target surface area. An additional criterion for a conductor having a fixed charge is that the separation between the field meter head and the conductor should not be so small that the capacitance of the latter is significantly increased and its potential decreased ($V = Q/C$).

The surface potential on a plane nonconductor, such as a charged web or belt, can be "measured" using normal field meter techniques. The readings are very useful for comparative purposes but are not quantitative. The measured electric field is proportional to the surface charge density as discussed in (3) above. However, the surface potential is not constant (C-2.5.6). As the meter head approaches the charged non-conductor, the local potential decreases owing to the capacitance introduced by the meter head. For small separations, an option for precise measurements is to use a field meter head that has the capability of being "floated" at elevated potential. A control loop is used to adjust the head potential until a zero field reading is obtained. This approach measures the electrostatic space potential in the vicinity of the sensing head [217].

3-5.3. *Potential*

In systems with very small charging currents and/or low capacitance, it is not possible to measure voltage with common volt-ohm meters, which have a typical impedance of 20–50 kΩ/V. This is because the instrument will rapidly drain away the charge. The charge on a conductor, hence its potential, decreases to about 37% of its initial value in one RC time constant (Eqs. 2-3.7 and 2-3.8). To allow time for a measurement to be taken the exponent of leakage resistance R introduced by the meter must be sufficiently large to offset the negative exponent of capacitance expressed in Farads. As noted in A-4-1.3, the capacitances of many common objects lie in the range 10 to 1000 pF (10^{-11}–10^{-9} F). For example, to obtain a time constant of 100 s at 10 pF the meter should have a resistance of the order 10^{13} Ω.

Potentials can be measured with zero current drain using electric field meters, since these are noncontact devices (3-5.2). These instruments must generally be used for measurements on nonconductive surfaces or conductors having small capacitance. In the case of a nonconductive surface the "potential" measured is of qualitative value only and represents only a proportionality to the surface charge density. A high impedance contacting voltmeter can be used for conductors having sufficient capacitance for a small current drain to be acceptable. The contacting voltmeters to be described introduce negligible current drain if properly maintained. The most significant current drain can sometimes be that needed to charge the meter plus its input cable.

3-5.3.1. *Electrometer*

The input impedance of electrometers is of the order 10^{14} Ω. Electrometers may be equipped with either capacitance or resistance voltage dividers to enable their voltmeter input range (typically 10 to 100 V) to be increased to 30 kV or more. The capacitance divider comprises an isolated target electrode held by plastic at the top of a metal cup and normally shielded by a grounded plate. The insulated rim of the inverted cup is placed on the charged surface (voltage V_1) and after momentarily grounding the target electrode the grounded shield plate is removed. Immediately a voltage (V_2) is induced on the target electrode. For fixed geometry, the capacitance C_1 of the charged surface to the target electrode is constant and the capacitance C_2 of the target electrode to ground, comprising cables and electrometer input, is also constant. The capacitances C_1 and C_2 comprise a voltage divider for which $V_1/V_2 = C_2/C_1$. Commercial devices are designed with division ratios of 1000:1 and if the cup is well maintained the leakage resistance is determined principally by the input impedance of the electrometer. Resistance voltage dividers with the same division ratio may also be used but these have a relatively low input resistance of the order 10^{11} Ω so are unsuitable for applications where this would represent an intolerable current drain, such as space potentials and measurements on nononductors. It is impractical to significantly increase the input resistance owing to impracticalities of maintaining higher value resistors in the divider probe without deterioration and calibration drift, especially at high voltage.

3-5.3.2. *Electrostatic Voltmeter*

These are mechanical devices comprising parallel capacitor plates. The outer plate or stator is held at ground potential. The unknown potential is applied to the highly insulated inner plate or rotor, which is diamond or sapphire-pivoted to move against a torsion force provided by a wire. From Coulomb's Law the instantaneous torque developed is directly proportional to the square of the instantaneous terminal voltage. Hence the deflection of the rotor as indicated by an analogue device gives a direct indication of voltage. Typically the instrument has a suitable calibrated scale which is illuminated by a light beam reflected from a mirror carried by the torsion wire. These voltmeters measure both DC and RMS alternating voltages regardless of wave shape between a few hundred volts and up to about 100 kV. Usually the low limit is about 1000 volts. Internal capacitance is usually very small, typically 8–10 pF for instruments designed to operate above 5 kV, although capacitance is greater for instruments designed for smaller voltages. Current drain due to capacitance is likely to be determined by the high voltage cables used. Leakage resistance is typically 10^{15} Ω for a well maintained instrument and is usually greater than 10^{14} Ω. At 10 kV and 10^{15} Ω the current drain of 10 pA should be negligible although if leakage resistance falls below 10^{14} Ω this will often no longer be the case. Leakage resistance can be measured by momentarily charging the device and measuring the voltage at increasing times. The time constant $\tau = RC$ seconds, where C is the voltmeter capacitance. As shown in 2-3 the time constant is the time taken for the reading to fall to half the initial value, divided by 0.693.

3-5.3.3. *Field Meter Voltmeter*

By placing a grounded electric field meter head in fixed relation to a highly insulated conductor having small capacitance and large radius of curvature, such as a metal sphere, and calibrating the system for indicated field against known potentials on the conductor, a voltmeter of extremely high input impedance is obtained. The range of such instruments is linear from a few volts to more than 100 kV. The geometry is normally arranged so that the output from the field meter in volts per meter calibrates to directly read the potential in volts on the conductor. To avoid perturbation by changes in capacitance with objects and people near the instrument, the device is often screened by placing it inside a grounded Faraday cage or metal box. To use the device, the conductor is connected to the source of potential via a suitable cable. If the source of potential is a nonconductor, the probe may comprise a fine corona wire or carbon fiber. The voltmeter may achieve extremely high input resistance up to about 10^{16} Ω, this being limited only by the insulation used to isolate the fixed conductor. Hence there is minimal current leakage and circuit loading. To minimize current losses in charging the voltmeter, the capacitance of the fixed conductor and cable must be minimized. In practice the high voltage input cable usually has a larger capacitance than the 10 pF typical for the isolated voltmeter. Elimination of corona losses is required for high voltage measurement, hence the need for large radii of curvature and highly insulated connections.

3-5.3.4. *Space Potential from Field Meter*

If a field meter sensing head is inserted into a large volume containing a space charge, for example a silo containing charged dust or a tank containing charged mist, the field indication is directly related to the local potential that would exist at the measurement point in the absence of the field meter. Provided the field meter head remains many tens of head diameters away from nearby grounded surfaces the local potential is related to the indicated field by $V = kE$, where k is a constant related to the sensing head diameter. In many cases the instrument manufacturer will supply the constant of proportionality. An example given in [51] is that a 9-cm-diameter field meter head indicated 11 kV/m per 1 kV of local potential.

3-5.4. Ignition Energy

The minimum ignition energy (MIE) required to ignite flammable mixtures by a spark is highly dependent on the test conditions, especially in the case of dusts. Measured MIE values give only a guide to the probability of ignition by other types of discharge such as brush types. In such cases the maximum "effective energy" of the discharge must be considered instead (2-5.1 and 3-8).

3-5.4.1. Ignition Energy of Gases

The MIE of gases varies with fuel–oxidant composition (5-1.4). The lowest minimum ignition energy (LMIE) of most gases in air is typically less than 1 mJ and occurs close to (between 0.9 and 1.9 times) the stoichiometric concentration. Data are given in Appendix B. The LMIE corresponds to the most easily ignitable, or "optimum" gas composition determined using an optimized spark circuit. Measurement requires great care; stray capacitance must be minimized and fully accounted for, plus the capacitor discharge circuit must be carefully isolated from the power source. Test procedures are given in ASTM E 582.

Figure 3-5.4.1 was constructed from data in Table A-4-1.3. It shows that hydrogen's LMIE is found for small diameter spark electrodes using a circuit having a small capacitance and hence a small RC time constant. The same general trend is also found for methane and is due principally to reduced

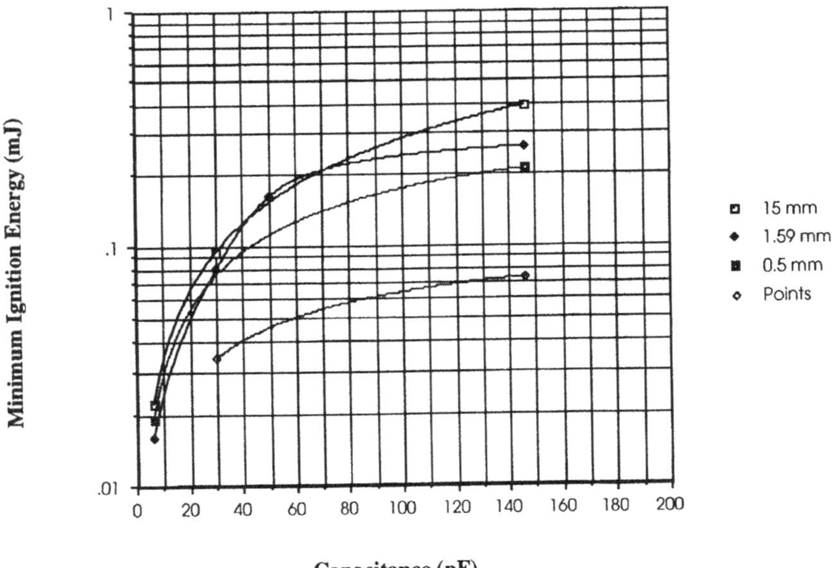

FIGURE 3-5.4.1. Effect of electrode diameter and circuit capacitance on MIE of hydrogen in air.

heat losses to the electrodes. For any given MIE value, the need to store energy at low capacitance means that a correspondingly greater voltage is needed. Figure 3-5.4.1 implies that pointed electrodes with a very small circuit capacitance (~5 pF) would allow ignition at less than hydrogen's reported LMIE of 0.016 mJ. Using pointed electrodes it is not possible to maintain the correspondingly greater voltage in a fixed spark circuit without corona discharge, although such a spark might occur in a practical situation involving a moving electrode. It follows that gas LMIEs are highly dependent on test technique and values given in Appendix B are probably overestimated in many cases (acetone?). Conversely, if the gas LMIE value is used to analyze the static hazard of some arbitrary spark-gap circuit, the result will normally be conservative even if the gas composition is optimum.

3-5.4.2. *Ignition Energy of Powder Suspensions*

See 6-1.2. Unlike gases, dusts cannot be tested as quiescent, homogeneous mixtures with an oxidant. Routine test methods involve dispersion of a dust sample in a test bomb of at least 1.2 L volume by a blast of air, followed by spark ignition. The method of dust dispersion varies, but more important are the methods used to time the spark occurrence after dust injection, the characteristics of the spark circuit and the methods used for calculating the energy released in the spark gap. Spark characteristics, particularly spark duration, affect gas and mist MIE to some extent. However, the effect on dust MIE is pronounced. Owing to variability in dispersion, the concentration and size distribution of the suspensions sampled by successive sparks is far from constant. Consequently the MIE varies from test to test [9]. This introduces an ignition probability, which means that repetitive tests must be carried out to ensure an optimum mixture is tested. The MIE generally decreases as more repetitions are carried out. Since the spark "samples" a small volume of the suspension, in the case of polydispersed or frangible dusts there is a probability that a spark will "sample" a system of particles of unrepresentatively small diameter. Thus, given enough tests, the MIE should decrease toward that of the smallest particles present. It is found that the probability of ignition is a simple continuous function of ignition energy. Therefore a probability plot could in principle be generated based on a large number of tests. From such a plot it would be possible to determine the MIE at a given ignition probability. However, this approach requires an enormous amount of work even to obtain a single probability [222]. For routine testing it is desirable to minimize the number of test variables and repetitions involved. Usually a predetermined dispersion system, spark gap geometry and spark circuit are adopted after initial optimization, and only dust concentration and stored energy are varied for each dust tested. Ignition tests are usually made on sub-200 mesh material or finer material representative of that present in a particular item of process equipment.

Test methods reporting the decrease of stored energy during discharge of a capacitor use the relation

$$W = 0.5C(V_1^2 - V_2^2) \qquad (3\text{-}5.4.2)$$

where the spark energy (W) in Joules is calculated from the total circuit capacitance (C) in Farads, the spark gap breakdown voltage (V_1) and the voltage remaining on the capacitor following the spark (V_2). It is important that total circuit capacitance includes all stray capacitance due to energized cables, electrodes and components such as nonisolated voltmeters. The capacitance of typical unscreened high voltage (≤40 kV) cables is up to about 40 pF per meter depending on location with respect to conductive surfaces. The capacitance of coaxially screened high voltage (~60 kV) cables is approximately 110 pF per meter. The error introduced by neglect of stray capacitance is greatest for high voltage, low capacitance circuits such as that described in BS 5958 [2], especially when measuring small MIEs. It is often overlooked that the energy stored at 15–20 kV is equivalent to 1–2 mJ for each 10 pF of stray capacitance. Since high voltage MIE tests use voltages up to at least 20 kV, stray capacitance can lead to an equipment resolution of several mJ. Underestimation of stored energy due to neglect of stray capacitance might explain some very low dust MIE values reported in the literature.

Test methods reporting the energy dissipated in the spark gap, calculated by integrating current and voltage across the spark gap with respect to time, yield smaller MIE values than methods reporting stored energy. This is because energy is always lost in the external circuit. Also, methods employing a component such as a 1–100 kΩ resistor or 1–2 mH inductance in the discharge circuit (to extend the spark duration) generally result in smaller MIE values than are obtained from capacitive circuits. This is related to the time required for most dusts to vaporize and mix with air before combustion can occur. There are two reasons why such methods are unsuitable for routine "hazard evaluation" test work. First, they are relatively sophisticated and require optimization both of technique and test variables. Second, they can give results that overstate the ignition hazard. For example, the MIE of 75–90 μm polystyrene dust was reported as 0.3 mJ [210]. The test method used a Hartmann tube with sharpened, 1-mm-diameter electrodes spaced 6 mm apart. A 1000 pF capacitor was discharged through a 6800 Ω series resistor and the voltage was measured before and after the discharge using a voltage divider arrangement. It was stated that of the 24.5 mJ stored energy, only 0.31 mJ was actually discharged in the spark. 0.3 mJ is within the range of typical gas LMIE values and the result suggests that +200 mesh polystyrene can be ignited approximately as easily as styrene monomer. This conclusion makes little sense in hazard evaluation. If stray capacitance or a voltage measurement error was not responsible for the result, support is given to the hypothesis that commonly reported gas MIE values are themselves too high (3-5.4.1). Also the use of "routine" dust MIE data is supported for hazard evaluation.

Routine methods of obtaining MIE are to either (1) construct the roughly parabolic curve of ignition energy against dust concentration, then read off the minimum value or (2) test only at preset ignition energies (such as a sequence where the energy triples at each step, viz, 1, 3, 10, 30, 100, etc., mJ) and report the MIE as falling between any two steps after 10 successive ignition failures at the lower energy level. For example, tests conducted under the auspices of ASTM found that the MIE of *Lycopodium* is about 20 mJ via the first method and 10–30 mJ via the second. A performance-based ASTM standard procedure for dust MIE testing was close to completion at the end of 1998.

3-5.4.3. Ignition Energy of Hybrid Mixtures

This can be measured by injecting dust into a premixed flammable gas–air mixture so that the final gas concentration is known. Provided the gas does not condense when pressurized it is possible, with appropriate precautions, to inject the dust using a pressurized gas–air mixture identical to the mixture in the test vessel. This avoids concentration gradients. Because it is a non-standard technique the possibility of estimating the hybrid mixture MIE (HMIE) should be considered (6-1.3).

3-5.5. Conductivity of Liquids

See 2-3.1. Typical laboratory conductivity meters have insufficient sensitivity to measure semiconductive and nonconductive liquids. In Appendix B, some tabulated conductivities appear as "<" suggesting that the instrument used was inappropriate. Some liquids listed as "conductive" might fall instead into the "semiconductive" category (e.g., cymene). For conductivities less than 100 pS/m especially, highly sensitive picoammeters are required to measure the small currents involved and great care is needed to avoid contamination of both the sample and the test cell. Several ASTM methods are available according to the conductivity range involved [143–146].

Conductivity decreases with decreased temperature to the extent that a measurement at room temperature will not represent a practical cold weather situation. The approximate variation of conductivity with temperature for distillate fuels such as gasoline and kerosene is [134]

$$\log_{10}(\kappa_2/\kappa_1) \cong \alpha(T_2 - T_1) \qquad (3\text{-}5.5)$$

where κ = conductivity at temperature T (°C)
α = constant(~0.015 /°C)

Hence if a laboratory measurement at 25°C yields a conductivity of 100 pS/m the same liquid at –10°C will have a conductivity of about 30 pS/m. The effects of low temperature combined with the elevated dielectric constants of many nonconductive chemicals support use of the 100 pS/m demarcation for nonconductive liquids (5-2.5) rather than the 50 pS/m demarcation used since the 1950s by the petroleum industry. For most hydrocarbons used as fuels, the dielectric constant is roughly 2 and a demarcation of 50 pS/m is adequate, provided the conductivity is determined at the lowest probable handling temperature.

3-5.6. *Resistivity of Solids*

See 2-3.1. Electrical conduction through solids takes place both through the bulk material and over the surface. In most cases surfaces have different physical and chemical properties than the bulk, for example due to contamination or moisture. Volume and surface resistivity can be separately measured for solid materials such as antistatic plastic sheet. Powders represent a special case since although both surface and bulk conduction occur, their contributions cannot be individually measured and the "volume" or "bulk" resistivity of a powder includes surface effects.

3-5.6.1. *Volume Resistivity of Powders*

Commercial devices have been produced especially for powders, the principal use being to assess suitability of candidate powders for electrostatically applied coatings. One design comprises a squat, cylindrical acrylic cell with parallel plate electrodes on opposite inner faces. The cell is opened, the powder added, and after tamping down the powder the cell is closed so that powder contacts both inner electrodes. The cell is then placed into the measuring device, which comprises a stabilized voltage supply (high or low switched voltage according to the resistivity of the powder) and sensitive current meter. The cell may be designed with unity cell constant so measured resistance across the cell equals volume resistivity of the powder.

3-5.6.2. Volume and Surface Resistivity of Solids

Test methods are described in ASTM D991 and ASTM D257. Commercial test equipment is available to facilitate electrometer measurements. In the case of sheet materials such as plastic packaging for sensitive electronic components, the Electronic Industry Association's Standard No.541 and the EOS/ESD Association's Standard S11.11 might be referred to when defining a "static dissipative" material (see "Glossary" for Surface Resistivity). There are several shortcomings in using resistivity values to assess the static dissipative properties of sheet materials. It is generally more appropriate to use measured charge decay time constants. A contemporary test method uses a fast-response field meter to measure the decay of surface voltage from a charged patch on the test specimen. For an initial voltage of 100 V a "static dissipative" material for use around semiconductors is defined as having a decay time constant of 10–500 ms. The 10-ms limit generally avoids static discharges while the 500-ms limit is considered adequately short where the charging rate is consistent with manual rubbing actions [218].

3-5.7. Resistance

For clean metal surfaces in the absence of stray currents a resistance up to about 1 MΩ can be measured by almost any quality volt-ohm meter while resistances up to 10^{13} Ω can be measured using electrometers. However, many such instruments use an internal voltage source of only 1.5 V. Where stray currents may occur due to cathodic protection or galvanic effects, etc., the voltage generated between the measurement points may offset the reading or in some cases give negative resistance readings. Also, since conductors are often painted or oxidized, a surface resistance may be present that cannot be overcome by a small source voltage. In both cases the problem can be avoided by using a higher voltage test instrument. Megohm meters (such as "meggers") use a source voltage of about 500 V and this is usually adequate to cause spark breakdown of thin, nonconductive surface layers of paint or rust. The upper ranges of some megohm meters approach that of the electrometer. It should be ensured that a megohm meter is not a source of ignition in flammable atmospheres.

A special type of resistance tester operates on an induction principle and does not require disconnection of the ground circuit being tested. A current is induced to flow in the ground circuit and the resistance is calculated automatically from the current sensed by a detector. The hand-held device is clamped around the ground cable and gives a direct reading of ground resistance. One commercial device measures six ranges of ground resistance between 1.0 and 1200 Ω.

3-6. Direct Observation of Discharges

In some cases it is possible to simply turn out the lights or wait until it is dark and directly observe the type, frequency and location of static discharges. This process can be facilitated using remote observation via a video camera with high sensitivity to low light levels. Where the discharges are faint and remote from the observation point, as might be the case with brush discharges inside equipment, the sensitivity can be increased using an image intensifier, or night vision device, which typically gives light amplification of 5000 to 50,000 times and can be attached directly to a video camera [12]. Where real time observation is not needed, another technique involves the use of high speed film and a camera on open shutter [121]. Both techniques have been used to record bulking brush discharges during filling of powder silos [12,121].

3-7. Radio Frequency Detection of Discharges

Since a closed, grounded metal tank behaves as a Faraday Cage, no externally generated electromagnetic wave can enter. Any radio frequency (RF) detected by an antenna inside the tank must therefore have originated inside the tank. If there are no other sources of RF inside the tank, it follows that a detected RF signal must be due to a static discharge. Not only has this technique been used to detect spark discharges and trigger devices such as camera flashguns, but (with limited success) attempts have also been made to estimate the energy of the discharges [98]. If commercially available RF receivers are used they must be AM rather than FM types, since the latter employ clipper circuits for suppression of "static" radio interference; in [67,160] the technique is described using a wideband AM receiver operating in the frequency range 450-480 kHz.

3-8. Measuring the Effective Energy of Nonspark Discharges

See 2-5.1. The effective energy is measured by subjecting a flammable gas mixture of known spark MIE to a large number of discharges so that an ignition frequency for ignition can be found for the conditions used. By recording ignition frequency at a series of increasing test gas MIEs the ignition frequency can be extrapolated to the required "zero" approximation, such as an ignition frequency of 5%. In the case of brush discharges from charged plastic, variable conditions include electrode diameter and gap length, surface charge density, gas mixture MIE and some characteristic dimension of the surface, for example, the diameter of a charged disc or pipe. The usual objective of gas ignition testing is to determine the largest credible effective energy of a nonspark discharge under specified conditions, as discussed in 2-5.1. Gas mixture composition and resulting MIE can be varied continuously above the LMIE value, either by varying the fuel concentration in air or by diluting the air with nitrogen. As noted in 2-6.2.1, it is possible that the gas composition most frequently ignited by a brush discharge might not correspond to the composition most easily ignited by spark.

3-8.1. Gas Composition

As shown in Figure 2-6.2.1 the MIE is extremely sensitive to gas composition, especially as the flammable limits are approached. The MIE of the gas mixture used for testing must be accurately known, so not only must superior mixing and gas analysis be used, but there must also be no change of composition during the test, for example, due to absorption of flammable gas by an oil or divergent gas flow into air from a probe. Further, the MIE-composition curve for the gas must be known, which in itself is a sophisticated test, especially at low MIE values (3-5.4.1). To minimize the rate at which MIE varies around a specified composition, a test gas with large flammable range, such as ethylene, should be selected in preference to a gas with small flammable range, such as propane (6-7.3.1). The MIE of gases having a wide flammable range is relatively insensitive to minor changes in gas concentration close to the optimum mixture. This is illustrated by Figure 5-1.4a which contrasts benzene (small flammable range) with ethylene oxide (large flammable range). Ethylene's small LMIE and large flammable range offsets experimental limitations when addressing the susceptibility of common fuel and solvent gases to a particular type of discharge. It is a simple asphyxiant with a LMIE roughly midway between methane and hydrogen.

3-8.1.1. Discharges from Solids

The test gas must be of uniform and known composition. This generally requires on-line gas analysis if flammable mixtures are not supplied from a suitable reservoir. If concentration gradients are created in the surrounding air, errors can be introduced by releasing the test gas stream from a perforated probe doubling as an electrode. The maximum effective energy of a brush discharge from a given charged surface is obtained using the largest diameter electrode capable of producing gas breakdown. However, if the electrode diameter is too large for the application no discharges will occur (C-2.5.3). It is therefore necessary to optimize the diameter using a series of electrodes. For brush discharge assessment, electrode diameters in the range 7 to 70 mm have been used (2-6.2). For discharges that only occur in large containers, such as the bulking brush, direct gas ignition tests are typically impractical owing both to the scale of the experiment and the difficulty in supplying a uniform test gas whose MIE is in the required range (10–50 mJ). In any case, a gas ignition test would not address the question of whether dust having a spark MIE equal to the measured effective energy could also be ignited (2-5.1.1).

3-8.1.2. Discharges from Nonconductive Liquids

Further to the considerations in 3-8.1.1, charged non-conductive liquids introduce two additional difficulties. First they are mobile; the charged liquid surface moves by electrostatic attraction toward the grounded electrode and charge is transferred by spraying plus, in some cases, via complete bridging of the gap. Second, the fuel component of the test mixture is absorbed rapidly into most oils causing the composition to become leaner. These effects reduce the measured effective energy and might be additive.

Charge is transferred by two-way liquid spraying between the disrupted liquid surface and the electrode. In response to local electric field forces, liquid deposited on the electrode tends to form small "Taylor" cones which may promote corona discharges in addition to spraying charged liquid back to the surface [8,71]. If the gap length is less than about half the electrode diameter, a stable liquid bridge may form between the liquid surface and the electrode. The bridge is tapered in the middle but widens at the ends and can attain a large fraction of the electrode diameter. At small charging rates brush discharges are suppressed via conduction through the bridge, whereas at high charging rates they occur around the bridge causing the bridge to collapse. The absence of ignition in a large-scale experiment might be due to a reduction of effective energy or even complete suppression of discharges via liquid surface deformation at small charging rates. This might result in a calculated "minimum surface potential for ignition" that is instead related to some minimum charging rate beyond which stable bridges cannot form. In any case, a fixed electrode does not represent practical cases where a hose or sample thief approaches a liquid surface faster than the surface can deform in response to the changing electric field. Small scale tests or direct observation should be carried out to ensure that the selected combination of charging rate, electrode diameter and gap length do not support formation of stable liquid bridges.

Fuels such as diesel and kerosene readily absorb hydrocarbon vapors, the total uptake and absorption rate depending on both chemical and physical factors. If a soluble test gas is introduced above a charged test oil the concentration of flammable test gas therefore decreases with time. Liquid mist and spray produced by charged liquid increase the absorption rate relative to a quiescent liquid surface. As discussed in A-5-4, absorption could lead to an underestimation of test gas MIE near the liquid surface unless the rate of test gas introduction is sufficiently high to offset the rate of removal. Table 3-8.1.2 shows solubilities of a selection of gases in a mineral-based transformer oil at ambient temperature and pressure [200].

Oxygen is roughly twice as soluble as nitrogen. The solubility of aliphatic hydrocarbons increases with chain length for methane through butane. Considering that methane has a relatively small LMIE, large flammable range and small solubility, it offers several advantages over propane test gas which has typically been used in published studies. The table does not address absorption rate. As a reference, when the author inverted test tubes containing various pure gases over a pool of diesel oil, ethyl ether was completely absorbed within one minute and lower aliphatic hydrocarbons in 3–6 min. Since test gases typically contain only a few percent of fuel component, absorption can be a considerable source of error. During operations such as switch-loading (5-1.4.3) absorption of vapor initially in the tanker from a previous lading may move the composition either into or out of the flammable range, introducing an additional variable into the accident statistics.

TABLE 3-8.1.2. Solubility of Gases in a Mineral Oil

Gas	Solubility (% vol/vol)
Oxygen as air component	3
Nitrogen as air component	7
Nitrogen	8.6
Air	10
Oxygen	16
Methane	30
Propane	1900
Butane	2000

4

CONTROLLING ELECTROSTATIC HAZARDS

4-1. Bonding and Grounding

4-1.1. *Definitions*

Bonding is the process of connecting two or more conductive objects together by means of a conductor so that they are at the same potential as each other but not necessarily at the same potential as the earth. Grounding is a specific form of bonding in which one or more bonded, conductive objects is also connected to ground so that each is at the same potential as the earth. Hence, all grounded objects are at the same potential and static discharges cannot occur between them.

4-1.2. *Purpose of Bonding and Grounding*

These measures prevent hazardous potentials from developing on conductors at the greatest anticipated charging current. This does not require a short circuit since charging currents are typically less than 100 μA. Depending on the system, allowable resistance to ground may be 1 MΩ or more.

4-1.2.1. Bonding and Grounding versus Individual Grounding

A series of conductive components might be bonded together and the bonded system then grounded. Alternatively, each conductive component might be individually grounded. The first option will probably require a shorter total length of cable and a neater appearance. However, in certain cases it is preferable to ground each conductive component individually, since loss of a single bonding connection might otherwise result in isolation of a system of conductors having a large capacitance. This could greatly increase the ignition probability where the individual conductors have a small capacitance and are considered to present only a marginal ignition hazard. The consequences of lost bonding connections via such mechanisms as vibration, corrosion, and maintenance activities should be considered in the grounding system design.

4-1.3. Resistance to Ground

The resistance to ground should be sufficiently small to prevent spark ignition at the maximum anticipated charging current to the system. This can be achieved by ensuring either that the energy stored is less than the MIE or that the minimum ignition voltage cannot be attained (A-4-1.3). The necessary resistance depends not only on the flammable mixture but also on the electrical circuit.

Depending on the flammable mixture a source voltage less than 100 to 1000 V will normally rule out spark ignition in purely capacitive circuits. However, if the stored energy ≥MIE, ignition in an interrupted inductive circuit might occur at much smaller source voltages (A-4-1.3). The conservative approach is to ensure that the ground resistance is sufficiently small that the maximum stored energy is less than the MIE of the flammable mixture.

Since gases have smaller MIEs than dusts, they represent the more challenging case for grounding. The most challenging case is where powders and flammable gases are handled together in air, combining the high charging rates produced by powder flow with the small MIE of flammable gases. For simple capacitance sparks, the MIE of gases decreases as storage capacitance and electrode tip diameter are decreased (Figure 3-5.4.1). The required resistance to ground can be found by considering the worst credible case flammable mixture, storage capacitance and electrode diameter. As shown in A-4-1.3 a ground resistance criterion can be written in terms of MIE

$$R_L < I_c^{-1}(2W/C)^{0.5} \qquad (4\text{-}1.3)$$

where I_C = charging current (A), W = MIE (J), and C = capacitance (F).

A ground resistance $R_L < 1\ M\Omega$ is satisfactory in practical cases even for hydrogen, which has an unusually small MIE. For dusts and most other gases a 1-MΩ grounding criterion is conservative owing to their greater MIEs and up to 100-MΩ is often satisfactory. Where lower MIEs may be encountered, such as the handling of anesthetic gases in oxygen enriched atmospheres, grounding requirements are more stringent as given in NFPA 99.

Exception: The resistance to ground of any conductive object that might be struck by a surface discharge on bulked powder in a silo should be less than 1 kΩ (6-4.1.1). This is because the average current transferred to the object may be of the order 10 A; the normal "1 MΩ" criterion will not always prevent the object from creating sparks to adjacent grounded points. The same mechanism may occur in liquid tanks via brush discharges or surface streamers although flow rate limitation as discussed in Chapter 5 should normally avoid such events.

4-1.4. Bonding and Grounding Systems

The 23 line drawings in [130] are very helpful for most routine bonding and grounding applications and for many years have been reproduced in NPCA 803 [6]. The following two sections cover general nonroutine applications. Chapters 5 and 6 of this book address specific grounding applications and special cases where electrical continuity may be unexpectedly lost.

4-1.5. Ground Rods

These devices are typically used to establish long-term or permanent grounding points for such purposes as lightning protection (NFPA 780). The general purpose of a ground rod is to achieve good electrical contact with the earth, so it should be sufficiently long to penetrate to a depth where the soil retains moisture. As described in Section 3-13.1 of NFPA 780 "Lightning Protection Systems," ground rods "shall be not less than ½ in. diameter and 8 ft long. Rods shall be copper-clad steel, solid copper, hot dipped galvanized steel or stainless steel. Rods shall be free of paint or other nonconductive coatings." For lightning protection and similar applications involving large currents, rod installation should follow NFPA 780. However, if a temporary ground rod is needed exclusively for dissipating static electricity, such as during emergency response, the installation of this type of grounding rod is usually impractical. In most cases an identifiable grounding point such as a buried metal water pipe or grounded equipment will already exist in the area. Where no such grounding point exists and there is a significant risk of static ignition, a temporary ground rod might be considered. The requirements for a temporary "static" ground rod are much less stringent than for a "lightning protection" or other high current ground rod. The maximum "static" current to be conducted to the earth is less than 1 mA. Based on formulas given in [216] it can be shown that a resistance to ground of the order 20 kΩ or less should be attainable using a temporary grounding rod approximately 1.5 m long, even when driven into dry sand. A typical resistance in soil is less than 1000 Ω. A temporary grounding rod used exclusively for static grounding might consist of a ½- to ¾-in.-diameter stainless steel spike that is driven into the soil using a sledge hammer. In most cases it is unnecessary to measure the resistance to ground for individual rod installations or to use water or brine to decrease the resistance. If a standard installation procedure is needed this should be developed using ground resistance data appropriate for the rod diameter, driven depth, type of soil and (if applicable) the use of water/brine additions. Ground resistance for the anticipated range of soil types should properly be determined using the Wenner four electrode method in which an AC potential is impressed on the outside pair of four linear electrodes and the potential drop is measured across the inner two (ASTM G57). However, an approximate method for temporary installations might be to measure the resistance between the temporary grounding rod and a second rod driven 1–2 m away, using either a megohm meter or high input resistance ohm meter. If a low voltage ohm meter is used, offsets due to stray currents might be ruled out by reversing the cable polarity and repeating the measurement. Provided the resistance between the rods is less than about 50 kΩ the temporary grounding rod will provide adequate grounding of small "static" currents. A temporary "static" ground rod is inadequate for dissipating large currents and should be removed after use to prevent misapplication.

4-1.6. Grounding and Cathodic Protection

Cathodic protection systems, using either a sacrificial galvanic anode or impressed current, require electrical isolation of the protected equipment from the rest of the plant. This special requirement can be defeated by general application of bonding and grounding, for example by installing a jumper cable across a flange containing an insulating gasket. More subtle defeats include cases where instruments or powered equipment are used on the protected equipment and share a common ground with the rest of the plant. The situation is especially complicated where tanks are protected by secondary containment systems involving plastic linings. A corrosion engineer or cathodic protection specialist should be consulted when tanks and other equipment are protected by such systems. To ground a tank that is isolated from the rest of the plant, one or more permanent ground rods (4-1.5) can be used which are not connected to the general plant grounding network. These are typically spaced at intervals around the tank. An alternative for lightning protection is a "Kirk Cell," formerly known as a "dual zinc system." This is essentially a capacitor, typically installed across an insulating flange, that becomes conductive in the event of a high voltage transient. Newer solid state devices are also available for these applications.

4-2. Control of Charge Relaxation

The rate of charge relaxation may in some cases be increased by increasing the conductivity either of the charged medium or of the surroundings. Examples include

- increasing the conductivity of a nonconductive oil using a commercial antistatic additive or miscible conductive liquid,
- increasing the conductivity of a plastic bag with a topical antistatic agent, and
- using an ionizing blower to increase air conductivity above nonconductive plastic foil.

4-2.1. Increase of Conductivity

This method may be applied to volume conductivity of liquids and solids, and surface conductivity of solids. For liquids see 5-2.6. The method is not usually applied to powders for quality and practical reasons, although moisture can have a significant effect (4-2.2.1). For solids such as sheet materials and manufactured items the conductivity can be increased using bulk additives such as carbon black, typically at several wt% loadings, or surface treatment with topical antistatic agents. The latter have the advantage of a wide variety of formulae which might be found to accommodate regulatory agency requirements for products used in food and drugs. They are typically designed to reduce surface resistivity from 10^{14} to 10^{16} Ω to the range 10^{8} to 10^{10} Ω. However, topical antistats may be removed relatively easily and are not always suitable where long term action is needed. Also, the action of most topical antistats relies on hygroscopicity and a certain minimum relative humidity is required; such agents often lose effectiveness where the relative humidity is less than 10–20%. Organometallic-based topical antistats not requiring the presence of moisture and capable of surviving processing temperatures up to 300°C have been described [201] which might be suitable for some low humidity applications.

4-2.1.1. Effects of Increased Humidity

Increasing the relative humidity of the air to 45 to 60% can significantly reduce charging where rates of surface separation are relatively small and the surfaces have time to equilibrate with the moisture. Hydrophobic surfaces such as polyolefins are less affected although humidities above about 70% have been used with success. Personnel shocks in offices and similar situations may be mitigated by increasing the humidity, particularly during winter when indoor heating results in very low relative humidities. However the method is unreliable for preventing ignition of flammable atmospheres by people (4-3). Atmospheric humidity has no effect on liquid conductivity and may have little effect on the charging of solids, especially where rates of separation are high or where the solids are hydrophobic. Examples include high speed web processes described in NFPA 77.

Increased relative humidity increases the conductivity of powders which readily absorb moisture, such as agricultural products. This effect can be significant at quite low relative humidity values and is pronounced above 30%. Sorption of water has two additional effects that reduce ignition probability in such cases. First, the tendency to produce suspensions of fine dust is decreased and second, the ignition energy is increased (6-1.6). For hydrophobic powders the effects of humidity are complex and high humidity might actually increase the net charge-to-mass ratio by encouraging unipolar as opposed to bipolar particle charging (6-3.1.1). Provided the air is not supersaturated, the effect on ignition energy is minor owing to the small amount of energy required to evaporate superficial moisture.

4-2.2. Charge Neutralizers

These devices increase the conductivity of the air so that charge drains away to ground. There are several types of commercial device, of which the two distinct types are those requiring external power ("active types") and those that don't ("passive" types). Neutralizers relying on free ions are limited by ionic mean free paths of a few centimeters in air, so cannot be used to neutralize charge in large volumes such as powder silos. This limitation can be offset by transferring the charge to tiny particles which can carry the charge over large distances (4-2.2.1).

4-2.2.1. Active Neutralizers

These typically use AC corona discharges to produce ions of both polarities, often in combination with a blower to help move the ionized air to the charged surface. Owing to the larger current developed they have a longer range than passive neutralizers but are still limited by mean free path of the ions. One commercial device uses a convergent–divergent nozzle to create a cloud of minuscule water droplets from humidified air entering the throat.

An AC corona discharge in the throat leads to a cloud of charged droplets whose large momentum allows very long travel distances. Multiple devices for eliminating static discharges in powder silos have been tested [41] but the tests did not address typical flow rates for large capacity, dense phase pneumatic transfer operations.

Note: DC resistive-coupled and capacitance-coupled corona neutralizers have been shown capable of producing large pulsive discharges from the needle electrodes whose peak values could attain about 5 A, even though the average corona current was only a few microamperes (2-6.1.1). Ignition of stoichiometric ethylene in air was reported for positive coronas [39]. The pulsive discharges could be produced when a conductor, for example a hand or metal tool, approached the neutralizer needle. Effective ignition energy depended on the coupling resistance and capacitance of the high voltage discharge line. Incendive pulsive discharges could be controlled by increasing the coupling resistance depending on source voltage, and/or using a thyristor-based switching device.

4-2.2.2. Passive Neutralizers

A variety of passive neutralizers have been used or proposed depending on the application. All types rely on the creation of ions.

1. *Needle, String, and Tinsel Bar Types.* These are used to remove charge from flat surfaces in close proximity to the neutralizer, such as moving belts. Field intensification by conductors with sharp edges such as needles creates local corona discharge and ions of opposite polarity flow to the charged surface until the field at the sharp edges falls below the corona threshold. Complete neutralization is therefore impossible and always limited by the corona threshold of 0.5–3 kV. To improve discharging efficiency, radioactive isotopes have been used to create ionization below the corona threshold, almost always in combination with a passive neutralizing system since the practical size of radioactive source is limited. These devices are described in NFPA 77. Recently introduced “string” or “ionizing cord” is a low weight, low cost static neutralizer incorporating thousands of tiny conductive microfibers such as stainless steel. The string is grounded at one end. An advantage is that the string is nonabrasive and may be allowed to contact moving surfaces without damage [225].
2. *Silo Inlet Neutralizers.* Two types intended to reduce the intensity of bulking brush discharges in powder silos are described in [73]. The first is a pointed discharge rod mounted axially inside the filling pipe outlet so that it points upstream into the filling pipe. The second is a

plastic-weighted wire mounted axially within the filling pipe so that it hangs down into the powder heap. Both should be securely grounded and the diameter of the discharge rod or wire should be 1–3 mm [73]. There is little information on the success of such devices in large commercial equipment and problems could include mechanical failure exacerbated by drag from incoming and bulked powder, plus decreased effectiveness caused by accumulated powder and coatings on the metal surfaces.

3. *Discharging Rods.* These have been widely used to dissipate charge from open powder containers such as drums and tote bins and comprise a rigid, grounded rod which is inserted close to the bottom of the container before starting powder flow [12,15]. It is normally assumed that rods need to be cylindrical and to encourage corona discharge it has been recommended that the rod diameter be kept to less than 3 mm [73]. It is proposed instead that square section rod be used, giving a smaller effective radius of curvature while allowing a greater rod thickness and rigidity. For conductive powders in plastic or plastic-lined containers it is especially important to insert the discharging rod before starting powder flow since delayed insertion can produce sparks from the powder surface (6-2.2.1).

4-3. Control of Personnel Charging

Personnel shock and spark hazards are discussed in 2-7. The level of perception for bare skin is approximately 1 mJ while significantly greater energy levels can be discharged imperceptibly when holding a metal tool; hence, imperceptible sparks from people may ignite common flammable gas mixtures in the work environment. Stored energies up to 10 mJ are commonly attained by ungrounded people. The maximum effective energy of a spark from a person is probably in the 30–50 mJ range, although 25 mJ has been considered as a more practical maximum value [11].

4-3.1. Personnel Grounding

Grounding devices should be selected so that hazardous charge accumulation on people is prevented while the risk of electrocution is not increased. Even for an ignition sensitive gas such as hydrogen, a ground resistance up to 10 MΩ will prevent hazardous charge accumulation even if the charging current to the person is as high as 100 μA (A-4-1.3). Personnel grounding often involves maintaining a resistance of about 10–100 MΩ between the skin and the ground. To prevent the possibility of electrocution via the grounding device should an electrical power source be contacted, a minimum ground resistance of at least 100 kΩ is usually imposed to limit the current to much less than the AC "can't let go" half-percentile level of 6–9 mA [63]. A typical minimum resistance is 1 MΩ, which results in currents less than 1 mA.The simplest type of commercial device is a grounding bracelet with built-in resistor, typically giving a resistance to ground of about 1 MΩ. This has the greatest utility at fume hoods and other situations where limited mobility can be tolerated. A hood might be equipped with two external coiled grounding cords with removable cuff attachments, the latter being kept by individual users for hygiene reasons such as the avoidance of scabies transmission. Volt-ohm meters or commercial testers can be periodically used to check ground continuity according to the manufacturer's specified limits. Where mobility is essential, footwear and flooring must be made sufficiently conductive instead. Antistatic footwear with resistance in the range 1 to 1000 MΩ (typically 10 to 100 MΩ) is commercially available, but for personal hygiene reasons is normally purchased for individuals. Alternatives include conductive straps and overshoes, which are especially useful for visitors to an area where antistatic safety shoes are being worn. If the conductive straps fit around the heel, ground contact may be lost when the wearer is squatting down; a clean room fire was considered in part to have been caused by loss of grounding while an operator was replacing a filter element close to floor level. Whatever is selected must be worn correctly, tested regularly, and be used in conjunction with housekeeping procedures as required to ensure the floor does not interfere with continuity. Nonconductive powder, oils, or lacquers can easily create a nonconductive coating on floors which in some cases may be difficult to inspect visually. Ground testers for floor resistivity are commercially available and procedures are described in NFPA 99.

4-3.1.1. Practicalities of Personnel Grounding

Personnel grounding is typically a risk-based decision taking into account the likelihood of encountering a flammable atmosphere and the consequences of an ignition, which are both affected by the type of operation and degree of confinement. Sparks from people are most likely during heated indoor operations owing to low relative humidity. Indoor operations are also most amenable to providing personnel grounding systems, including maintaining housekeeping for antistatic or conductive flooring. Conversely, it might be impractical to provide assurance of personnel grounding for outside activities such as equipment maintenance and tank operations. For example, antistatic shoes and flooring might become disabled by contamination with nonconductive material. There are few accounts of ignitions from people in the open air. Operations involving tank openings present a significant risk although it is difficult to determine whether reported ignitions were due to static discharges from personnel or from equipment and containers being used.

4-3.2. *Clothing*

The use of clothing made from high resistivity synthetic fibers such as polyester should be avoided more because of their poor fire resistance and tendency to melt into the skin in response to even a small flash fire than because of static hazards. The use of fire resistant clothing is discussed in detail in [54]. There is usually a very small likelihood of ignition by grounded people due to any type of clothing having a reasonable fit, unless the clothing is removed. The crackling noise when removing overgarments such as sweaters is common experience, and is due to the production of corona-brush discharges. If articles of clothing are removed, ignition might occur as a direct result of separation of charges by the clothing, even if a person is grounded. Removal of garments such as coats and sweaters should always be done remote from flammable areas. The likelihood of ignition by ungrounded people can be increased by clothing having high resistivity. An example is the charging that occurs with respect to seat covers when getting out of a car or forklift truck. If clothing can be charged by rubbing there is a greater likelihood of ignition if an ungrounded person is wearing synthetic, rather than cotton or antistatic clothing. In critical areas such as clean rooms and explosives manufacturing, the use of antistatic clothing is frequently used in addition to grounding plus, in extreme cases, the banning of metal jewelry such as rings and wristwatches that could facilitate sparking. A more common reason for banning these items is where an area is slippery and a fall could result in the item being caught on an adventitious hook. For example, a finger may be torn off by a ring.

4-3.3. *Gloves*

These should ideally be conductive or antistatic so that handled items that are troublesome to individually ground, such as tools, are grounded through the grounded person. Also, where antistatic shoes are not worn, people have a high probability of being frequently grounded via the glove. In many cases suitable gloves are commercially available. However, in some cases industrial hygiene concerns dictate specific impermeable glove materials that may not be available in antistatic form. There is no general solution to this problem. In most cases, small conductive items such as tools have too small a capacitance to give an incendive spark. An exception is regions of very large space potential, such as in the vicinity of charged plastic sheet or when lowering a conductor into a large tank containing charged mist. In these identifiable cases, items can be individually grounded if suitable gloves are not available. Similar precautions are needed where gas mixtures of very small MIE, nominally less than 0.1 mJ, are involved.

4-4. Control of Flammable Atmospheres

General measures for controlling flammable atmospheres inside equipment are described in NFPA 69 "Explosion Prevention Systems." These measures do not necessarily prevent flammable atmospheres in some practical cases, such as when powders are added to an inerted vessel via an open manway. Where possible such cases are identified in the relevant sections of Chapters 5 and 6.

4-4.1. Liquid Nitrogen/Liquid Air Hazards

Liquid air has a boiling point less than that of liquid oxygen and high pressure liquid nitrogen (LN2), but air condenses easily on low pressure LN2 lines and systems containing liquid hydrogen or helium. Liquid air can, for example, condense in porous insulation on LN2 systems and form liquid pools. The first drops that condense and the pool that initially forms both contain about 50% oxygen [131]. As the liquid air evaporates, nitrogen boils off first and the pool becomes enriched with oxygen, approaching pure oxygen in the final stages. An explosion occurred in a chewing gum manufacturing process when LN2 used for equipment cooling caused condensation and enrichment of liquid air in the presence of flammable dust; the explosion involved a dust–air suspension with local ignition in an oxygen-enriched mixture. There is also a potential for concentrating liquid mixtures containing liquid or solid flammable materials, which may be explosive and shock sensitive [132]. Owing to the very low ignition energies of flammables in oxygen-enriched atmospheres, measures should be taken both to avoid the hazard and to mitigate static accumulation in areas where the phenomenon might occur.

5

FLAMMABLE LIQUIDS, VAPORS, AND GASES

5-1. Ignition Hazards of Liquid Vapor and Mist

5-1.1. Flammable Liquid

Liquids with closed-cup flash points less than 100°F have a recognized ability to form flammable vapor–air mixtures under ambient conditions and are commonly known as flammable liquids. While liquids with flash points ≥100°F usually require an elevated temperature to form flammable vapor–air mixtures and are sometimes known as combustible liquids, other effects such as flash point error, reduced ambient pressure, dispersion into fine droplets or some combination of these should be considered when using the reported flash point to assess flammability hazard. Some liquids with a low fire hazard, for example paints comprising mostly water, may be classed by NFPA 704 as combustible even though they can generate explosive vapor-air mixtures in closed containers at less than 100°F; the NFPA fire hazard rating can be as low as 1 which normally indicates a flash point ≥200°F. Similarly, some liquids with no measurable flash point may generate flammable vapor via degassing or slow decomposition in large containers, especially where the vapor space is small compared with the liquid volume. In such cases the vapor space may need flammability evaluation under actual use conditions. The use of a 100°F demarcation between flammable and combustible liquids is gradually being replaced by the UN/DOT 141°F (60.5°C) demarcation, which allows for elevated temperatures during transportation. The NFPA and DOT systems for flammable and combustible liquids can be found in the "Glossary" under "Class." These systems both assume "round number" demarcations at 140°F and 200°F, although the 140°F breakpoints are not consistent. NFPA Class IIIA combustible liquids have flash points ≥140°F while DOT combustible liquids have flash points >141°F.

5-1.1.1. Flash Point Test Error

Flash point does not in general represent the minimum temperature at which a pool of liquid will form a flammable atmosphere. Closed cup test methods involve downward flame propagation, which is more difficult than upward propagation, and the region where the test flame is introduced is normally fuel-lean relative to the liquid surface. Also, the volume of the closed cup test apparatus is too small to allow flame propagation of certain flammable vapors such as halogenated organics. Limitations of the flash point test are discussed in the Appendix to ASTM E 502, where it states, “If process or handling conditions dictate the usage of a flammable material at temperatures ranging upward from 5–10°C below the closed cup flash point, then a flammable vapor might be present above the liquid, and the potential hazard might be more precisely defined by determining such properties as temperature limit of flammability (Test Method E 1232) or flammable limit concentrations (Test Method E 681), or both, at the contemplated conditions.” While closed-cup flash points are usually lower than open cup values, this is not always the case as noted in the Appendix to ASTM E 502.

5-1.1.2. Liquid Temperature Error

The surface temperature of stagnant liquid in a heated or uninsulated tank may significantly exceed the bulk liquid temperature owing to heat transfer from the unwetted upper walls. An uninsulated tank might be heated by sunlight to about 60°C. Since vapor–liquid equilibrium is established at the vapor–liquid interface, an excessive liquid surface temperature can result in an elevated vapor concentration relative to the calculated value based on bulk liquid temperature. Vapor in the tank may be flammable even if the bulk liquid temperature is less than the reported flash point; an unrecognized hazard could result if the vapor is assumed to be below the lower flammable limit (LFL). Conversely, even where the tank vapor space is clearly above the upper flammable limit (UFL), an ignitable mixture will always exist in the vicinity of an open vent. Furthermore, vapor vented from large air-breathing field storage tanks may be at only 30–50% of theoretical saturation based on bulk liquid temperature. Hence a tank explosion hazard might be overlooked if the entire vapor space is assumed to be above the UFL (5-1.2).

5-1.1.3. Ambient Pressure Error

The equilibrium vapor pressure above a confined liquid depends only on temperature. The fraction of the total pressure exerted by vapor pressure determines the composition of the vapor–air mixture. Thus when the total pressure is reduced, for example at high elevations or in vacuum trucks, the vapor concentration in air increases. Since flash points are reported at a standard pressure of 760-mmHg an ambient pressure less than this value depresses the effective flash point. The observed flash point at pressure P (mmHg) can be estimated from the standard flash point (760 mmHg) using an adaptation of the correction formula given in ASTM E 502

$$\text{Observed flash point} = \text{standard flash point} - N\,(760 - P) \quad (5\text{-}1.1.3)$$

where constant N (0.03 for flash point in degrees Celsius and 0.06 for flash point in degrees Fahrenheit) should be considered only approximate.

5-1.1.4. Low Concentration Volatiles Error

Small concentrations of volatile components in a liquid mixture may accumulate in the vapor space of a container over time and appreciably reduce the flash point relative to the reported closed-cup value. This may be the result of degassing, chemical reaction or other mechanism. An example is bitumen [162]. Similarly, if a tank truck is not cleaned between deliveries of gasoline and a high flash point liquid such as kerosene or diesel oil, the mixture might generate a flammable atmosphere both in the truck tank and the receiving tank. Contamination at the thousand ppm level may create hazards (5-1.4.3 and 5-2.5.4). Solids containing upward of about 0.2 wt% flammable solvent need to be evaluated for flammable vapor formation in containers (6-1.3.2).

5-1.1.5. Safety Margin for Flash Point Application

In view of the above adverse effects a safety factor should be applied where flammability is assessed using flash point. For pure liquids in containers the vapor should be considered potentially flammable if the liquid temperature is upward of at least 5°C below the reported flash point. For mixtures whose composition is less certain, such as petroleum mixtures, the safety factor should be about 15°C relative to the flash point [55]. Where combinations of adverse effects are identified the safety factors should be increased accordingly. A simple but very conservative approach is to assume that all liquids having a flash point ≤141°F may produce a flammable atmosphere under some ambient conditions, even where no mist or froth production is involved. A more practical approach is to assume that liquids handled in air at least 5–15°C below their closed cup flash points will not present ignition risks unless

- they are handled in a manner that produces mist or froth,
- they contain small concentrations of volatile material that is not preserved during flash point sampling and testing, or
- they generate flammable gas via slow reactions.

5-1.2. Flammable Limits

The flammable range in air lies between the lower flammable limit (LFL) and upper flammable limit (UFL) as determined by test for the conditions of interest. Outside this range of compositions the gas or vapor cannot be ignited. Operation at less than the LFL is often considered to be safer than operation at above the UFL, particularly for atmospheric storage tanks. Even if liquid in a tank rapidly generates sufficient vapor for operation above the UFL, flammable mixtures may occur around tank openings such as sampling ports, and the flammable range may be traversed inside the tank during start-up or other operational condition. However, there are many cases where operation above the UFL is essential, such as storage of high vapor pressure liquids, or has other advantages, such as some vent collection header applications [168]. Alternatively, vessel atmospheres can be rendered nonflammable using inert gas as described in NFPA 69. This technique reduces the oxygen concentration below the Limiting Oxygen Concentration (LOC) which is the minimum oxygen concentration required to sustain combustion. Inerting is usually ineffective near tank openings, especially in cases where solids additions occur and entrain air. Also, for

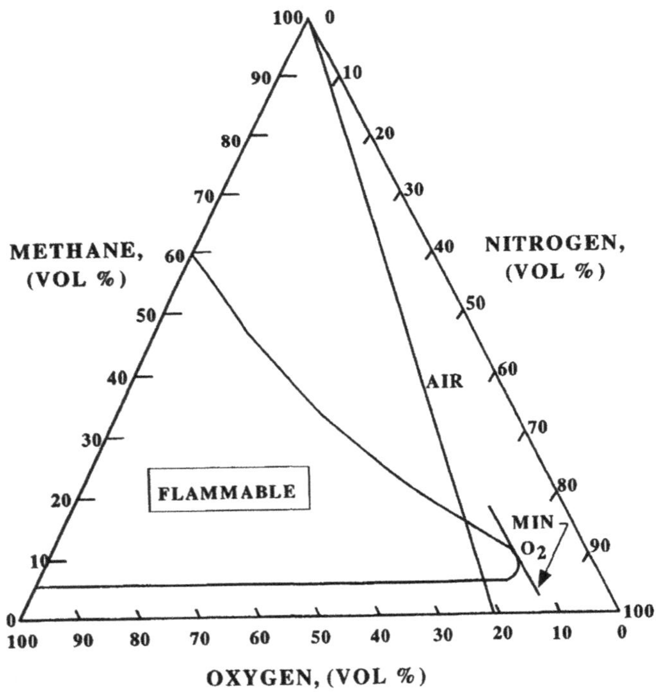

FIGURE 5-1.2. Flammability triangular diagram for methane + oxygen + nitrogen.

storage tanks, inert gas supply must be adequate to compensate for air in-breathing during tank emptying or changes in temperature.

5-1.2.1. *Effect of Diluents on Flammability*

On the air line shown in Figure 5-1.2 the flammable range of methane in air is 5–15 vol%. On the methane leg of the triangular diagram, the flammable range in oxygen is read as 5–60 vol%. At the nose of the flammability envelope the minimum oxygen concentration (min O_2) that will give a flammable mixture is 12 vol%. This is the limiting oxygen concentration (LOC), which is a singularity that occurs at fuel concentrations marginally above the LFL. At greater fuel concentrations the LOC becomes increasingly conservative in relation to the actual flammable envelope. A fuel gas such as methane may be added to increase the total fuel concentration and increase the margin of safety relative to the flammable envelope [168]. The exact locations of flammable limits depend on conditions such as temperature and pressure, plus the diluent concerned. The diluent is often nitrogen, but can be any gas that does not contribute to combustion (carbon dioxide, rare gas, water vapor, etc.) or some mixture of these. It is usually found that diluents with larger heat capacities decrease the flammable range and increase the ignition energy although the effects may be complex [142, 168–171].

5-1.3. *Liquid Mist*

Provided a flammable or combustible liquid is sufficiently dispersed in air to propagate a flame, the flash point of the liquid is irrelevant in determining ignitability of the mist. Even at very low liquid temperatures, frozen liquid droplets may burn in the same manner as a dust cloud. Ease of ignition and rate of combustion of a liquid mist both increase as droplet size decreases. Depending on liquid volatility, droplets with a diameter less than 10–40 μm typically vaporize and ignite ahead of a flame front and their overall combustion behavior is similar to that of a vapor (5-1.4.4). Since liquid mist production by shear processes also generates static electricity, it is good engineering practice to avoid splashing and other operations that generate mist inside equipment. Charging of ungrounded conductors by charged mist is discussed in 5-6 and elsewhere.

Apart from deliberate spraying through small orifices, flammable, finely divided mist is most likely to be created by rapid cooling of hot vapor or rapid expansion of pressurized vapor. These form the basis of numerous experimental techniques for creating dense, almost monodispersed mists less than 10 μm diameter. Condensation of hot oil vapor has commonly been used to produce ~0.6-μm mist found optimum for smoke screens [180]. Depending on cloud thickness, droplet size, and illumination, mists typically become optically dense at much less than their LFLs.

During processes such as splashing, the production of easily ignited mists of combustible liquids such as diesel oil is unlikely unless a froth layer is produced. This is because the liquid concentration in the mist will normally be less than the LFL. The LFL varies with the liquid and decreases as droplet size is increased. A typical LFL range for an oil is about 50 mg/L at 5 μm decreasing to 10 mg/L above 80 μm [179]. These concentrations are optically dense, especially for finer, more easily ignitable mists. However, there is adequate liquid in a froth layer to create a flammable mist of fine droplets (5-1.3.1). If ignition occurs, the flame can propagate into coarser mist above the froth layer. Subsequent combustion may involve all of the suspended mist in the container, augmented by droplets and vapor derived from any incoming liquid stream.

5-1.3.1. Froth or Foam

Aeration of a liquid, for example via splashing, forms gas bubbles which increase the liquid interfacial area at which charging occurs. In addition to increasing the charge generation rate, aeration may create a relatively long lasting froth which in turn is a source of very fine mist as it bursts. Burst rate is greatly increased locally by a static spark and the froth is completely disrupted ahead of any flame. Froths, like mists, can be ignited at less than the flash point of the liquid. Another feature of froths (see below) is that the two-phase system has a volume conductivity less than that of the continuous liquid phase, so may both accumulate a surface charge and hinder charge dissipation from any isolated conductive object in the froth layer. Such conductive objects are not restricted to obvious items such as soda cans. Other possibilities include metal debris buoyed up from the bottom of the tank, ice or hydrates derived from the hose or tank surfaces and isolated patches of liquid product or water.

If a froth is produced by degassing following a sudden decrease in pressure, previously dissolved gases will determine the combustion behavior of the froth. Owing to the very high solubility of light aliphatic hydrocarbons in most oils the appearance of flammable gas is expected in such cases. Oxygen is typically more soluble than nitrogen (Table 3-8.1.2), so the froth can additionally be enriched with oxygen during degassing.

In 1965 a powerful explosion and ensuing fire occurred while gravity loading kerosene to a barge under conditions producing excessive bubbling, foaming and turbulence [111]. The explosion occurred at much less than the liquid flash point of 110–130°F and the only credible mechanism was static ignition of mist inside the tank compartment. Kerosene is a nonconductive liquid. Since it was established that free water plus a large quantity of air flowed into the barge tank from the 10 in. supply line, charge accumulation was predictable (5-3.1.1). However, liquids do not need to be nonconductive to accumulate charge. In December 1987, semiconductive butyl acrylate liquid with a flash point of 49°C (120°F) ignited in a properly grounded tank car at about 15°C [30]. The ester was introduced at high velocity via a thrust neutralizer comprising four-way horizontal flow diversion at the foot of the filling lance. Flow rate was 1450 gpm through 6-in. pipe followed by a partly inserted 4-in. dip pipe. When an account of the incident was published it was believed that flammable vapor had been present due to a previous lading of methyl methacrylate [30]. However, in January 1989, an almost identical fire occurred in a dedicated butyl acrylate tank car, suggesting that both incidents were due to ignition of liquid mist at well below the liquid flash point. A possible explanation for both incidents involves high charging due to shear in the thrust neutralizer, production of a liquid froth, and static discharge from charged liquid or other material supported by the froth layer. While the conductivity of the ester involved is typically about 3000 pS/m the effective conductivity of a froth layer might be 1–2 orders of magnitude less [UCC unpublished]. The company involved subsequently changed to 105J cars with fixed dip pipes.

If froth is produced by decomposition of an oxidizing agent such as hydrogen peroxide, the oxygen-enriched froth may be flammable and burn rapidly even in a completely inerted vessel (5-9.6.1).

5-1.4. Minimum Ignition Energy (MIE)

While flammability exists throughout the flammable range the energy required to ignite fuel–air mixtures varies by orders of magnitude as the composition is changed. At the flammable limits the ignition energy by definition increases toward infinity. Flammable limit measurement typically involves electrical energies of the order 10 J. A minimum value known as the lowest minimum ignition energy (LMIE) or simply as the minimum ignition energy (MIE) is exhibited approximately half-way between the flammable limits [8]. The LMIE for most flammable vapors in air is between 0.1 and 1 mJ (Appendix B). Since the energy available from most electrostatic discharges is small (typically less than 10 mJ) relative to the energy used to establish flammable limits, the probability of ignition is usually strongly related to the probability of generating the most easily ignitable mixture. Liquids close to their flash points, or to their flammable limits, are generally not susceptible to weak static discharges. Brush discharges are particularly ineffective close to the UFL (Figure 2-6.2.1). The MIE is increased by a reduction in the oxygen concentration relative to air, for example by partial nitrogen inerting. However, this effect may not be pronounced until the oxygen concentration in the oxidant is decreased below about 16 vol%. This is based on propane–air–nitrogen mixture MIE data in [57]; in order to increase the MIE of propane from 0.25 to 1.0 mJ, an additional 23 vol% nitrogen needed to be added to air, corresponding to an oxygen reduction from about 21 to 16 vol% in the oxidant mixture.

Figure 5-1.4a shows how the MIE varies with composition for ethylene oxide and benzene mixtures in air. The former has a very low LMIE and extremely wide flammable range. In the presence of a static ignition source having an effective energy of 1.0 mJ, ethylene oxide may be ignited over a very much wider range of compositions than benzene. For an ignition source of 0.15 mJ energy, no possible benzene–air mixture can be ignited while a wide range of ethylene oxide–air mixtures remains susceptible to ignition. A particularly hazardous feature of gases having a small LMIE and wide flammable range, including acetylene, hydrogen and ethylene, is that the MIE remains small over a large range of compositions. There is a correspondingly greater probability of an easily ignitable mixture coinciding in

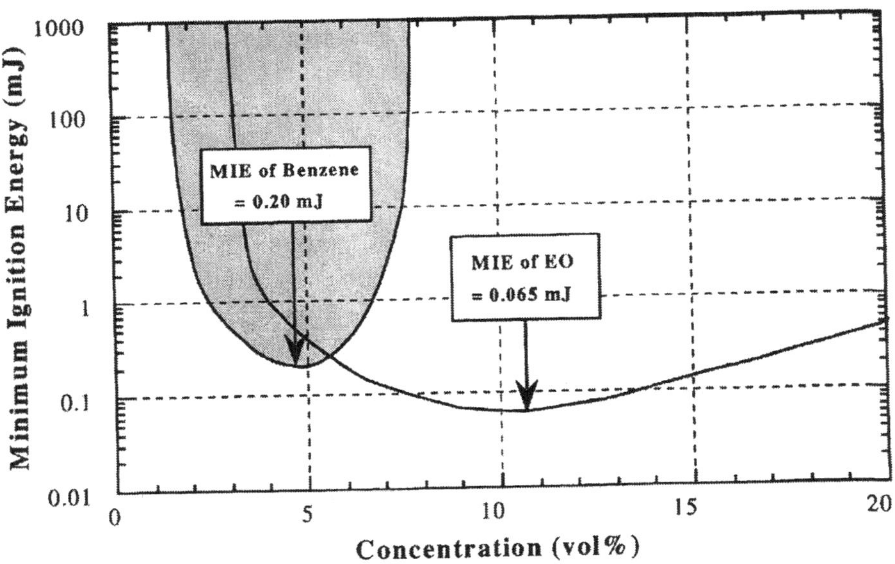

FIGURE 5-1.4a. MIE Curves for benzene and ethylene oxide in air.

time and space with a static ignition source. Appendix B provides a tabulation of LMIE values for a number of flammable gases and vapors.

Elevated temperature, pressure and oxidant concentration relative to air may significantly reduce MIE (5-9.6 and 5-9.7). Conversely, reduced pressure elevates MIE. Figure 5-1.4b, developed from data in [142], shows how propane's MIE varies with both absolute pressure and oxygen concentration in an oxygen–nitrogen oxidant mixture. The propane concentration is optimized in the range 5–15 vol% depending on oxygen concentration. In air (~21% oxygen) a pressure reduction from 1 atm to 0.2 atm increases the MIE from 0.25 mJ to nearly 6 mJ, a factor of about 25. The significance of this is that under many vacuum conditions such as in vacuum truck cargo tanks the MIE of typical hydrocarbons exceeds the maximum effective energy of brush discharges. As oxygen in the oxidant mixture is increased from 21% to 100% the MIE falls by a factor of about 100 at all pressures in the range examined.

5-1.4.1. Vapor Pressure

The vapor pressure (VP) can be used in conjunction with a measured LFL to estimate flash point. When this is done it is usually found that the calculated flash point is less than the measured value owing to limitations in the flash point test technique (5-1.1.1). Conversely, the LFL can be estimated only approximately from the flash point. The VP can also be used to replace the

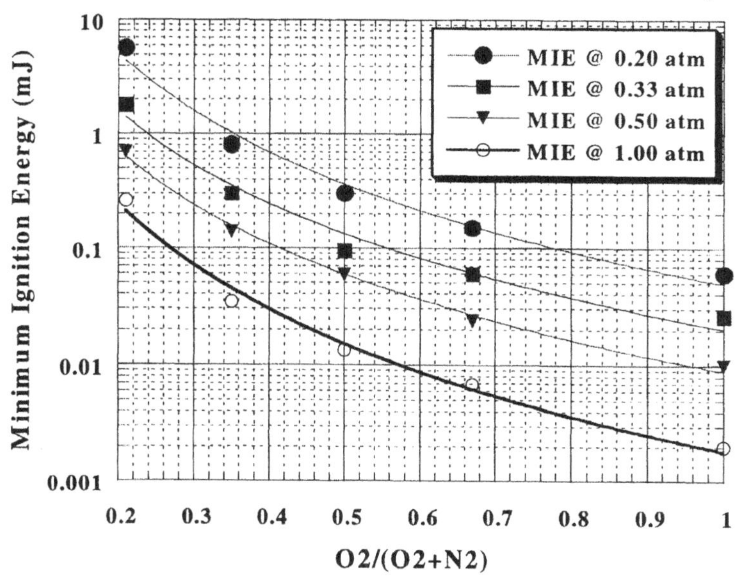

FIGURE 5-1.4b. Effect of pressure on MIE of propane in various oxygen–nitrogen mixtures.

concentration axis in Figure 5-1.4a with the corresponding temperatures required to generate these concentrations. In this way one can determine the equilibrium liquid temperature at which vapor ignition is most probable, corresponding to generation of the LMIE composition (5-1.4). For many aliphatic and aromatic hydrocarbons this composition is approximately equal to the average of the LFL and UFL, although this rule does not generally apply to other chemicals [8]. For vapor–liquid equilibrium, benzene generates its LMIE vapor-air mixture at about 7°C and toluene at about 26°C (Table 5-1.4.1). For operations conducted close to room temperature, toluene is more prone to ignition from low energy static discharges than benzene [8]. This is because benzene's equilibrium vapor concentration exceeds the UFL at temperatures above about 16°C. However, at colder temperatures of 0–5°C benzene is more prone to ignition, since toluene vapor is close to or less than its LFL. Included in Table 5-1.4.1 is an example of a VM&P naphtha (vergaellungsmittel petrolether). Since VM&P is primarily a mixture of various paraffins and naphthenes, plus minor aromatics such as xylene, various formulations can be readily characterized only by their closed-cup flash points (9°C in this case) and flammable limits (1–6% in this case). However, in the context of Table 5-1.4.1 it can be seen that this nonconductive liquid should be prone to static ignition at ambient temperature and pressure. This inference is supported by the incidence of tank truck loading fires. Petroleum naphthas having elevated flash points compared

TABLE 5-1.4.1 Approximate Vapor–Liquid Equilibrium Temperatures for LFL, LMIE, and UFL Vapor Concentrations at Atmospheric Pressure

Liquid	LFL (%)	LFL (°C)	LMIE (%)	LMIE (°C)	UFL (%)	UFL (°C)
Diethyl ether	1.9	–46	5.1	–28	36.0	9
n-Hexane	1.1	–25	3.8	–5	7.5	5
Cyclohexane	1.3	–18	3.8	4	8.0	15
Benzene	1.3	–12	4.7	7	7.8	16
n-Heptane	1.0	–4	3.4	14	6.7	26
Methanol	6.0	8	14.7	25	36.0	42
Toluene	1.1	4	4.1	26	7.1	37
(VM&P naphtha)	~1	~9	3.5–4.0 (est)	20–30 (est)	~6	Unknown
Ethyl benzene	0.8	15	3.8 (est)	48	6.7	58
Styrene, xylene(s)	0.9–1.1	23–28	3.8 (est)	~50	6.7–7.0	60–65

with motor gasoline are far more likely to be within the flammable range during tank loading.

The assumption of vapor–liquid equilibrium is an oversimplification for tank filling. The vapor concentration is maximum near the liquid surface and concentration gradients exist throughout the vapor space depending on the rates of diffusive and convective mixing. As discussed in 5-2.3.1, an undesirable side-effect of splash filling can be to dilute vapor near the liquid surface with entrained air. Consider a volatile, nonconductive liquid such as hexane, whose vapor concentration near the liquid surface normally exceeds the UFL. Air entrainment and turbulence introduced by splash filling may result in mixtures close to the liquid surface that are instead close to the LMIE composition. If splash filling is due to a partly inserted dip pipe equipped with a tee, positive brush discharges from the liquid surface to the tee might complete the fire triangle.

5-1.4.2. High, Intermediate, and Low Vapor Pressure Liquids

a. *High Vapor Pressure Liquids.* These are defined in API 2003 as having a Reid Vapor pressure exceeding 4.5 psia (31 kPa). At normal handling temperatures, rapid evaporation of such liquids during tank loading minimizes the duration of a flammable atmosphere above hydrocarbon fuels and the UFL is soon exceeded. In most cases these liquids are loaded into grounded tanks without prior inerting of the vapor space. Inerting might be used in cases where the liquid has an unusually high UFL. Note that air quality mandates are steadily resulting in production of less volatile gasolines with greater associated flammability hazards (5-2.6).

b. *Intermediate Vapor Pressure Liquids.* These are defined in API 2003 as having a Reid Vapor Pressure less than 4.5 psia (31 kPa) and a closed cup flash point less than 100°F (37.8°C). They are most likely to generate flammable vapor mixtures in vessels at ordinary temperatures. Although graphical methods have been widely used to estimate whether liquids are likely to generate flammable atmospheres at various temperatures, based on their Reid vapor pressures, these were originally derived for petroleum fuel mixtures and do not generally apply to other flammable liquids. Errors at the upper limit are significant for pure hydrocarbons such as ethyl benzene and cyclohexane, and can be large for oxygen-containing chemicals such as ethers.

c. *Low Vapor Pressure Liquids.* These are NFPA Class II and III combustible liquids (closed-cup flash point ≥100°F) and will generate flammable atmospheres only if handled at elevated temperature, suspended as a mist or subject to slow vapor evolution. However, static electricity generated during handling may ignite flammable vapor present from previous operations (5-1.4.3).

5-1.4.3. Switch Loading

This is the practice of loading liquid, typically a low vapor pressure liquid, into a vessel containing a flammable or fuel-rich atmosphere from a previous use. A flammable atmosphere may form during loading or may persist throughout loading. The worst case is for the atmosphere to remain at its most easily ignitable composition throughout loading, which might occur if there is negligible scrubbing of the vapor by the liquid being loaded (3-8.1.2). The worst case situation is similar to loading a pure liquid such as toluene whose vapor pressure at normal loading temperatures frequently results in an easily ignitable composition throughout loading. Thus, switch loading may be more hazardous than loading an intermediate vapor pressure liquid whose equilibrium vapor composition either traverses the flammable range or does not attain the most easily ignitable composition during loading. Switch loading has been the cause of numerous fires. Since the operation is no more hazardous than loading a pure liquid such as toluene having the same electrical properties, the root cause of these fires must be either failure to follow procedures appropriate for flammable liquid loading (such as limitation of flow rates) or unusually long relaxation times associated with the low vapor pressure liquid. This is the case with viscous nonconductive liquids such as some lubricating oils, which may require inerted conditions for switch loading (5-2.5.4).

Note: Section A-5-6.3 of the 1996 edition of NFPA 30 (reiterated in NFPA 385) contains erroneous information about switch loading.

> When a tank is emptied of Class I liquid, there is left a mixture of vapor and air, which can be, and often is, within the flammable range. When such a tank is refilled with a Class I liquid, any charge that reaches the tank shell will be bled off by the required bond wire. Also, there will be no flammable mixture at the surface of the rising oil level because the Class I liquid produces at its surface a mixture too rich to be ignitable.

Toluene is a Class I liquid (see "Glossary"). Not only is it unlikely to produce a rich mixture at its surface at ambient temperature, but the mixture produced is often close to its most easily ignitable composition throughout loading. This common misconception about switch loading is based on the behavior of gasoline and does not apply to Class I liquids in general. It is impossible for switch loading to result in more easily ignitable mixtures than occur when loading a pure liquid such as toluene at a temperature corresponding to its most easily ignitable vapor composition (5-1.4.1). This oversight in NFPA 30 may be traced to its traditional and primary application to storage of hydrocarbon fuels such as gasoline.

NFPA 30 also fails to recommend flow rate restrictions except a slow start until the downspout is submerged. Section 5-4 of this book provides for restricted flow rates throughout filling; this should be applied wherever charge accumulation is possible due to low liquid conductivity and where flammable mixtures involving gas, mist or froth may be formed.

5-1.4.4. MIE of Liquid Mists

When a liquid is dispersed into droplets the surface area is increased, which enhances the rates of heat and mass transfer. For a particular liquid dispersed at constant concentration in air the MIE varies with approximately the cube of surface average droplet diameter, hence the MIE decreases by a factor of about 8 when the surface average diameter D is halved (A-5-1.4.4). Ease of ignition is greatly enhanced for finely divided mists with D less than about 20 μm, whose MIE approaches that of the vapor. Below 10 μm a high flash point liquid mist (tetrahydronaphthalene) was found to behave like vapor while above about 40 μm the droplets tended to burn individually [142]. Since liquid mists must partially evaporate and mix with air before they ignite, the ease with which a liquid evaporates also affects MIE (Figure 5-1.4.4).

5-1.5. Explosion Prevention Systems

Prevention of a flammable atmosphere may be accomplished using any of the alternatives presented in NFPA 69. In cases where fuel concentration cannot be limited, the most common technique (inerting) is to add a suitable inert gas such as nitrogen, so that the residual oxygen concentration is insufficient to support a flame. A safety factor is then applied. For most flammable gases and vapors this typically involves reducing the oxygen concentration to less than 5–8 vol% (see Chapter 2-7 of NFPA 69).

5-1.5.1. Expanded Metal and Plastic Foam Packing

Various expanded metal and foamed plastic matrices are used as flame arresters when packed into flammable vapor spaces such as the fuel tanks of racing cars and military attack aircraft. The techniques are not mentioned in NFPA 69 and are not used in passenger or cargo transport aircraft. Advanced packing designs have used flexible polyurethane foams having very low mass density and void fraction. Following a number of static fueling fires on military aircraft in the early 1970s, plastic foams were developed

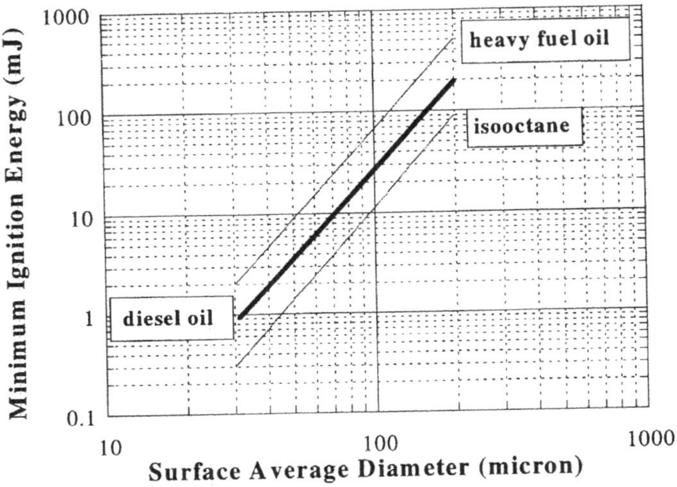

FIGURE 5-1.4.4. Schematic effect of droplet diameter on MIE of typical petroleum products.

having antistatic properties and various nozzle designs (such as piccolos) were investigated to minimize charging between the plastic foams and the jet fuels (notably JP-4) involved. Additionally, the fuel conductivity was increased to at least 200–600 pS/m. Information on this topic can be found in [112–118], of which the best introduction is [118]. There are probably valid applications in the CPI for passive protection systems involving packing and it might be argued that equipment such as packed columns already has this protection to some degree. However, packing that was not specifically designed and installed to prevent flame propagation at the conditions involved cannot be expected to act as a flame arrester, even though the extent of flame propagation inside the packed equipment is greatly reduced. Practical considerations for plastic foams include compatibility of liquids with the plastic foam, general quality concerns, and the cleaning, inspection and maintenance difficulties that would be introduced.

5-2. Generation and Relaxation (Loss) of Charge in Liquid Systems

5-2.1. Charge Generation

Ionic species present in liquids undergo adsorption at interfaces such that predominantly one sign of charge is more strongly bound at the contacted surface than the other. This results in a bound layer close to the surface farther from which is a diffuse layer having a net countercharge. This two-layer system having opposite net charges is referred to as an electrical double layer. For conductive liquids such as water the diffuse layer is only a few molecules thick but for nonconductive liquids such as light petroleum distillates the layer may be many millimeters thick. Processes producing surface shear separate the weakly held diffuse layer from the bound layer. For example, during flow of a liquid hydrocarbon through a pipe the diffuse layer is carried downstream and may result in accumulation of the net charge in a receiving tank. Charge separation occurs whenever shear effects occur at an interface so additional examples include splashing, spraying, and stirring. Where a large contact surface is available per unit volume the rate of charge generation is increased. Examples are filtration and agitation of two-phase systems such as liquid–gas, water–oil or powder–solvent suspensions.

5-2.2. Charge Density

The net charge (Coulombs, C) in a specified volume of liquid (cubic meters, m^3) gives charge per unit volume or charge density (C/m^3). Since flow in a pipe can be described by a flow velocity, and current is defined as the flow of charge per unit time, the flow of net charge carried along the pipe by flowing liquid represents a charging current or "streaming" current

$$\text{Charging current (C/s or Amperes)} = \text{Volume flow rate } (m^3/s) \times \text{Charge density } (C/m^3) \qquad (5\text{-}2.2)$$

5-2.3. Factors Influencing Charge Generation

See 2-1.1 and 2-3.1. The mechanism of charge generation is highly complex. For flow of liquid in pipes it can be shown that the charging current depends on liquid dielectric properties (electrical conductivity and dielectric constant), viscosity, and flow characteristics, involving factors such as flow velocity, pipe diameter and surface roughness. For equal flow characteristics, the electrical conductivity is the dominant factor affecting charging. This is most pronounced for nonconductive liquids ($\kappa < 10^2$ pS/m) since the conductivity of such liquids may vary by orders of magnitude depending on the concentration of trace contaminants. Trace contaminants have negligible effect on liquid dielectric constant and viscosity but have a dominant effect on conductivity. Owing to their much greater intrinsic conductivities, trace contaminants have increasingly less effect on semiconductive ($100 < \kappa < 10^4$ pS/m) and conductive ($\kappa > 10^4$ pS/m) liquids. In many systems such as long, grounded metal pipes it is found that the charge density attains a steady state at which the rate of charge generation is balanced by the rate of charge relaxation back to ground.

5-2.3.1. Splash Filling

This practice has been blamed for numerous fires and explosions, although often for the wrong reasons. It is frequently stated that static is generated when a liquid stream falls through the air and breaks into droplets. In practice this is a minor hazard when filling tanks, even where the liquid is nonconductive. Charge generated by stream break-up is almost entirely collected in the continuous liquid phase with negligible net charge accumulation. This has been verified by a large scale study which concluded "Splashing does not itself generate significant amounts of charge" [221]. Although charged mist forms, it has a very small charge density compared with the continuous phase and is typically neglected in tank filling models. The hazards of splash-filling are principally related to effects other than break-up of the incoming stream. These include the absence of electric field reduction by a metal dip pipe and the ability to accumulate excessive charge densities at the liquid surface via stratification. The static hazards of charged mist are normally significant only where operations such as sampling and gauging introduce ungrounded components into the tank. To allow dissipation of charged mist prior to gauging a "wait time" may be needed (5-5.2).

Splash filling may also affect flammability. Froth and mist created by splashing may be ignited at much less than the flash point of the bulk liquid (5-1.3.1). An opposite and previously unrecognized effect of splash filling is the associated dilution of vapor near the liquid surface via air entrainment and turbulence (5-1.4.1). When a tank is filled with a volatile liquid whose equilibrium vapor concentration exceeds the UFL, a large fraction of the vapor space remains flammable until sufficient evaporation has taken place. However, in the absence of turbulent mixing the UFL is very rapidly exceeded close to the evaporating liquid surface. Since ignition via brush discharges typically occurs within 10 cm of the surface, air dilution may create an ignition hazard above liquids that otherwise would quickly generate local vapor concentrations close to or above their UFLs. The probability of ignition of volatile liquids such as naphtha may therefore be increased by splash filling as opposed to bottom filling. The effect of vapor concentration on the frequency of ignition by brush discharges is discussed in 2-6.2.1.

5-2.4. Charge Relaxation

See 2-3. Provided κ is above ~2 pS/m, charged hydrocarbons such as gasoline contained within grounded, conductive containers or pipe lose charge to ground at a first order (Ohmic) rate governed by the following equations

$$\tau = \varepsilon_r \varepsilon_0 / \kappa \qquad (2\text{-}3.2)$$

$$dQ/dt = -Q/\tau \quad (2\text{-}3.6)$$

$$Q_t = Q_0 \exp(-t/\tau) \quad (2\text{-}3.7)$$

The relaxation time (τ) is a parameter governing the rate at which charge is lost from a liquid, accounting for both its electrical conductivity (κ) and permittivity ($\varepsilon_r\varepsilon_0$). The integrated rate equation (2-3.7) shows that 36.7% of the initial charge remains after one relaxation time (when $t = \tau$), 13.5% remains after two relaxation times, 5.0% remains after three relaxation times and 0.67% remains after five relaxation times. By providing three or more relaxation times between filling a tank and carrying out tank sampling or other hazardous operation, the charge introduced by the filling process will usually have decayed 95% or more. As discussed later, certain exceptions to ideal behavior can occur.

Case 1. Ungrounded Conductive Tank. When an ungrounded tank is filled, the potential on the tank at any time t is proportional to the total accumulated charge (A-4-1.3)

$$V_t = I_C R_L\{1 - \exp(-t/R_L C)\} \quad (5\text{-}2.4.1)$$

where I_C = charging current (A), R_L = leakage resistance from tank to ground (Ω), and C = tank capacitance (F).

The tank potential is independent of the relaxation time of the liquid, since the tank acts as a Faraday pail (3-5.1.2).

Case 2. Grounded Conductive Tank. It is sometimes possible to accumulate hazardous potentials on nonconductive liquid surfaces in grounded tanks. Since the ignition hazard depends on surface potential, which is directly proportional to the charge density in the tank, the relaxation time of the liquid is a critical parameter when modeling the variation of surface potential with time. It is typically assumed that the liquid is uniformly charged and in contact with grounded metal, even at the free surface. It was shown in 2-3 that the rate of charge loss from charged liquid is proportional to the charge remaining but is independent of the shape of grounded boundary. A charge balance on the liquid volume can therefore be written

$$dQ/dt = I_C - Q/\tau \quad (5\text{-}2.4.2)$$

For constant charging current, and boundary conditions $Q = 0$ when $t = 0$, the total charge is

$$Q_t = I_C\tau\{1 - \exp(-t/\tau)\} \quad (5\text{-}2.4.3)$$

This equation was originally given for liquid flowing through a large, full tank such that the liquid level was in contact with grounded walls over the entire boundary and the current loss via flow out of the tank could be neglected [55]. If both sides of Eq. (5-2.4.3) are divided by the volume filled at time t, one arrives at the time-dependent equation for charge density $Q_V(t)$ typically used in tank filling models [22]

$$Q_V(t) = (I_C t / V_t)\{1 - \exp(-t/\tau)\} \qquad (5\text{-}2.4.4)$$

where V_t = volume (m^3) filled at time t (s).

The assumption of constant charge density implies a mixing time that is small compared with the relaxation time. In reality the inlet jet always contains a greater charge density than the surrounding liquid in the tank and the free surface always contains excess charge. Also, charged liquid entering tanks with large *L/D* ratios such as tank trucks and rail cars is unlikely to be well-mixed at the ends. Conditions such as splash filling or partly inserted dip pipes during overhead loading, and undeflected jet penetration to the surface during bottom loading, can cause excessive accumulation of charge at the free surface resulting in elevated surface potentials and possibly ignition.

5-2.4.1. Hyperbolic Relaxation

If all liquids lost charge according to the Ohmic equations in 5-2.4, those with very low conductivities would have relaxation times of hours. However in the 1960s, Bustin showed that nonviscous petroleum distillates with conductivities less than about 2 pS/m relax charge much faster than predicted using the Ohmic equations [24]. Later analysis of Bustin's model showed that the time taken for high charge densities (100–5000 μC/m^3) to decrease to safe values (<20–30 μC/m^3) is roughly independent of initial charge density [9]. The charge dissipation time is approximately 90 s for oils with dielectric constant of 2 and 180 s for chemicals with a dielectric constant of 4 (A-5-2.4.1). Where liquid conductivities are less than about 2 pS/m, filters should ideally be placed at least 100 s residence time upstream (5-2.4.2). In Appendix B, any nonconductive liquid whose typical conductivity is less than 2 pS/m is assigned a nominal 100 s dissipation time rather than a relaxation time. For short filling times, such as drumming, this behavior is comparable to a that of an Ohmic liquid whose conductivity is 0.5 pS/m [8].

The hyperbolic relaxation equation (A-5-2.4.1a) contains charge carrier mobility as a variable, which should be sensitive to oil viscosity. This is found to be the case for some viscous nonconductive liquids. These have much slower rates of charge dissipation equivalent to an Ohmic liquid whose conductivity is 0.02 pS/m (5-2.5.4).

There are no available data to establish whether nonconductive, low viscosity chemical products such as ethyl ether similarly display hyperbolic relaxation below about 2 pS/m, or even whether this phenomenon is a practical reality for such liquids. Should Ohmic relaxation behavior continue to much less than 0.5 pS/m the risk of static accumulation would be enhanced compared with petroleum distillates.

5-2.4.2. *Residence Time Requirements*

With reference to Figure 5-3.5, a relaxation volume should be provided between a charge generator such as a microfilter and a tank so that the residence time is sufficient for charge to relax to safe levels. As discussed in 5-3.5 and A-5-2.4.1, the residence time for nonconductive liquids that relax charge Ohmically should normally be three relaxation times or a default value of 30 s. This criterion is often applied to all nonconductive liquids. In principle however, liquids with conductivities less than 2 pS/m that relax charge hyperbolically require about 100 seconds of residence time to achieve an equivalent reduction in charge density. Where practical, 100 s of residence time should therefore be provided for liquids whose conductivity is typically less than 2 pS/m. An exception is viscous nonconductive liquids, whose required residence time may be impractically long (5-2.5.4). Unless antistatic additives (5-2.6) can be used, inerting may be necessary if flammable atmospheres are present, for example, if switch loading is carried out.

5-2.5. *Classification of Liquids based on Conductivity*

The conductivities of flammable and combustible liquids vary from about 10^{-4} pS/m to 10^{10} pS/m, that is, by about 14 orders of magnitude. Dielectric constants usually vary over 1–2 orders of magnitude in the range 2–40; the higher values are generally exhibited by polar molecules which also have higher conductivities. Since relaxation behavior is primarily governed by conductivity this property can be used to assign any of three group classifications conductive, semiconductive, and nonconductive. These are useful in classifying the static dissipative characteristics of liquids for different charging situations. While the rate of charge relaxation generally increases with increased conductivity, the rate of charge generation such as during flow through pipes and filters often maximizes at the low end of the semiconductive range (about 100 pS/m).

Since conductivity is highly sensitive to sample purity and temperature, group demarcations can be given only to within an order of magnitude. It should always be considered that conductivity under practical conditions might be less than measured in the laboratory. Appendix B contains a listing of conductivities and relaxation times for different liquids. ASTM test methods are discussed in 3-5.5.

5-2.5.1. Conductive Liquids

Liquids having conductivities greater than 10^4 pS/m, such as acetone, are considered conductive. Their relaxation times are typically less than about 1 ms and for this reason they tend not to accumulate charge except where handling conditions isolate them from ground. These conditions include complete isolation in the form of a droplet suspension in air, partial isolation by suspension in another liquid, and containment in plastic or other highly resistive container. A separated, nonconductive second phase in the form of a froth (5-1.3.1) or skim layer (5-4.1.5) may isolate supported patches of charged liquid. Conductive liquids are most prone to induction charging by plastic containers and are sufficiently conductive to lose much of the induced charge in the form of a spark.

5-2.5.2. Semiconductive Liquids

Liquids having conductivities between 10^2 pS/m and 10^4 pS/m such as butyl acetate are considered semiconductive. Their relaxation times are usually in the range 2–500 ms. For this reason they tend not to accumulate charge except where charging rates are high and/or grounding is impeded, such as flow through rubber hose or end-of-line polishing filters, or where they are effectively isolated from ground (5-2.5.1), or where they are partially isolated by an underlying froth layer (5-1.3.1). Spark discharges may be possible from the more conductive of these liquids.

5-2.5.3. Nonconductive Liquids

Liquids having conductivities less than 10^2 pS/m such as purified toluene, diethyl ether and most low-sulfur diesel oils are nonconductive. Their relaxation times are generally above 180 ms and often greater than 1 s. Their dielectric constants are typically 2–6 and they are highly susceptible to conductivity variation due to trace contamination. In many cases, nonconductive liquids have negligible intrinsic conductivity and conduction is due principally to contaminants. For example, although commercial *n*-heptane typically has a conductivity of about 1 pS/m or more, after passing the liquid through silica gel to remove impurities its measured conductivity is less than 0.03 pS/m. This behavior is used to establish zero for conductivity testing according to ASTM D 3114. It follows that the conductivity of such liquids is extrinsic and may be drastically reduced by processes such as distillation. Nonconductive liquids are most prone to accumulate static and in some cases may generate hazardous potentials in grounded metallic containers.

For many years the petroleum industry has defined nonconductive liquids as having conductivities less than 50 pS/m. A higher value of 100 pS/m is used here to address the higher dielectric constants of certain flammable chemicals in relation to petroleum products. For example the dielectric constant of ethyl ether is 4.6 versus 2.3 for benzene; from Eq. (2-3.2), ethyl ether therefore has the same relaxation time at a conductivity of 100 pS/m as benzene at a conductivity of 50 pS/m. It is the relaxation time, not the conductivity alone, that determines the rate of loss of charge; hence the same logic that makes 50 pS/m appropriate for identifying nonconductive hydrocarbons makes 100 pS/m appropriate for identifying nonconductive chemical products.

Corona and brush discharges are observed from charged nonconductive liquids rather than sparks. Since only partial discharge is possible, induction charging from charged plastic containers is not a significant hazard. Relaxation (or dissipation) times are typically less than 100 s although certain viscous nonconductive liquids with viscosity above 30–100 cS may display longer relaxation times of the order 1000 s (5-2.5.4). In Appendix B, low viscosity liquids with very low conductivity are considered to display a nonohmic dissipation time of 100 s. This is to address the observation that as conductivity is reduced to less than about 2 pS/m, the rate of charge relaxation is governed by a hyperbolic rather than Ohmic law (5-2.4.1). Slow settling of suspended material in tanks may increase the effective relaxation time by generating a settling potential (5-4.1.2).

5-2.5.4. Viscous Nonconductive Liquids

Some liquids having a kinematic viscosity above 30–100 cS can have very slow rates of charge dissipation, equivalent to a conductivity of about 0.02 pS/m and a relaxation time of about 1000 s. Up to 5000 s may be needed to dissipate charge. Filtration charging both in microfilters and strainers is typically greater than for nonviscous liquids. For viscous oils through 10-μm filters, charge densities up to 5000 $\mu C/m^3$ are common while values this high are uncommon for nonviscous oils, even in finer filters. Unusually large concentrations of antistatic additive may be required to render the liquids semiconductive. Since safe charge dissipation times may be several thousand seconds, it might be impractical to provide adequate residence time and inerting should be considered where flammable vapor is present. Fuel oils, crude oils and lubricating oils containing detergent additives typically have high conductivities and do not pose this hazard. Conversely, base lubricating oils, turbine fuels and transformer oils not containing additives can have viscosities upward of about 100 cS and conductivities of the order 0.02 pS/m [189]. Viscous silicon-based oils might have similar properties.

On February 11, 1999 a 375-m^3 atmospheric tank was filled with luboil product from a barge. The viscous product (3000 cS at 20°C) had a flash point of 194°C and was trace heated to 90°C to facilitate handling. Following transfer, the line was blown down using compressed air at which point ignition occurred and the tank roof was blown off, landing on the ground beside the tank. There were no injuries or spills and the fire rapidly extinguished itself. It was found that the oil contained an unexpected 0.3 wt% hexane impurity which boiled off and formed a flammable mixture in the tank head space. It is predictable that static was generated during high velocity two-phase flow in the pipe. However, as air bubbles subsequently rose through liquid in the large tank, they might have convected significant charge to the surface via shear of the electrical double layer formed at the air–liquid interface. The charge separation effect of rising air bubbles should be analogous to that of settling water droplets, which produce a settling potential (5-4.1.2). Both phenomena may lead to charge separation in large tanks containing nonconductive liquids. Ignition presumably followed the generation of large potentials at the liquid surface, causing either a surface streamer to the wall (2-6.6) or a positive brush discharge to a grounded projection (2-6.2). Although hexane vapor ignition caused this incident, an aerated froth might in principle ignite irrespective of flash point (5-1.3.1). The incident demonstrates the general hazard of blowing down lines with air, plus the increased potential for static accumulation in viscous nonconductive liquids.

5-2.5.5. Changes in Conductivity Caused by Freezing

Liquids may undergo a sudden and dramatic decrease in conductivity at their freezing points. This might cause unexpected static hazards. For example, the conductivity of biphenyl decreases about 4 orders of magnitude between the liquid phase (above 69°C) and solid phase (Appendix B). A static ignition occurred when biphenyl at 120°C was loaded into a tank containing a thick layer of solid biphenyl from previous operation [13]. Hot biphenyl is normally sufficiently conductive to rapidly dissipate charge when loaded into a grounded metal tank, but owing to the presence of the thick, insulating layer of solid biphenyl, charge was able to accumulate and a static discharge occurred from the liquid surface to the filling pipe.

5-2.6. Antistatic Additives

The conductivity of nonconductive liquids can be increased by adding part-per-million (ppm) quantities of commercial antistatic additives or percentage quantities of a suitable conductive liquid (5-4.5.2). Commercial antistatic additives were formerly used mostly in aviation fuels but are now being increasingly used for other distillate fuels such as gasoline, in part due to the impact of clean air regulations. To reduce emissions, the Reid vapor pressures of gasoline blends are trending downward and the risk of developing flammable mixtures in tanks is trending upward. Complex formulations were originally developed by Shell in the late 1950s. These took advantage of synergistic effects between various additive components [55]. Shell's additive ASA-3 has been described as "chromium salts of mono-dialkyl salicylic acids and calcium aerosol didecyl sulphosuccinate bound in a co-polymer of lauryl and stearyl methacrylate and methyl vinyl pyridine." Disadvantages of antistatic additives containing heavy metals include toxicity and degradation of performance with time. Metal-free formulations replaced ASA-3 in the 1990s when EPA fuel additive mandates essentially restricted the constituents to compounds of C,H,O,N and S. Apart from additives used primarily for benefits other than conductivity enhancement, the only important antistatic additives are Octel's Stadis 425 and 450 (sulphonic polymer mixes) and Baker-Petrolite's Tolad 3512 (acrylate copolymer mix). It must be ensured that commercial antistatic additives are present at the required concentration in target locations such as tanks and are not removed by water layers or equipment such as clay filters.

Commercial antistatic additives are increasingly being used in a wide variety of chemical operations, the primary use currently being for nonconductive aromatic solvents such as toluene and xylene. Such solvents may alternatively be rendered conductive by the addition of a large concentration of conductive liquid, usually described by the paint and coatings industry as an oxygenated solvent. This practice is common in liquid–solid mixers (5-4.5.2). Addition of conductive liquid is dependent upon process constraints and chemical compatibility since the required concentration of conductive liquid, such as an alcohol or ketone, may be of the order 20 vol%. Also, the resulting mixture must be evaluated for changes in flammability characteristics. It should be noted that 1–5 ppm of commercial antistatic additive might be as effective as 10–20% of conductive liquid additive. If the primary solvent has a flash point above the process operating temperature, the addition of a low flash point conductive liquid such as MEK will greatly increase the flammability hazard. In such cases the operation might be made inherently safer by using a commercial antistatic additive. It is necessary to thoroughly evaluate the application to ensure that conductivity remains adequately high during the operation and that there is no impact on product quality.

5-2.7. *Bonding and Grounding*

See Sections 4-1.3 (general), 5-3.1 (pipe), and 5-3.3.1 (hose), which also address special grounding cases such as swivel joints.

5-3. Flow in Pipe, Hose, and Tubing

5-3.1. *Metallic Piping Systems*

All-metallic piping systems present minimal static hazards since there should be no impediment to bonding and grounding. All parts of continuous all-metallic piping systems should have a resistance to ground of no greater than $10^3\ \Omega$ and in practice a resistance of $10\ \Omega$ might be adopted as the maximum expected value. Higher resistances might indicate poor electrical contact although this will depend on the overall system and the expected value ($<10^3\ \Omega$) should be established following installation. For bolted, flanged pipe neither paint on the flange faces nor thin plastic coatings used on nuts and bolts normally prevents bonding across the flanged coupling after proper torque has been applied. Jumper cables and star washers are not usually needed at flanges; plant operators have reported that star washers can interfere with application of proper torque. In addition to nonconductive items, connections in metallic pipelines such as grooved-and-shouldered joints and swivel joints may however need jumper cables to provide electrical continuity. When painted, slip flanges (lap joints) using nonconductive gaskets may also cause loss of continuity (Figure 6-3.2). This can be remedied using a conductive gasket, such as a flexible, graphite filled, spiral wound type, or a jumper cable across the joint. It should be ensured that bonding efforts do not compromise deliberately isolated sections of pipe such as where insulating flanges are used to avoid stray current arcs or where cathodic protection systems are employed (4-1.6).

5-3.1.1. *Charge Generation in Pipe and Hose*

The rate of charge generation in long, conductive pipe and hose is eventually balanced by the rate of charge dissipation back to ground via the wall. This results in a steady-state charging current or streaming current which in a given system varies with the liquid involved and increases with flow velocity. Charging currents generally increase with flow velocity, pipe diameter and surface roughness. Turbulence promoters such as elbows, tees, valves, and orifice plates create a transient increase in charging current which subsequently decays toward the steady-state value (5-3.7). Charging currents are greatly increased by any second phase that increases the interfacial surface area at which charging occurs. Examples include microfilters (5-3.5), fine strainers (5-3.5.1), suspended immiscible liquids and solids (5-3.6), and gas bubbles (5-1.3.1). In industrial situations, charging currents generally vary in the range 0.01–100 μA, the high value representing extreme cases such as microfiltration and the low value to slow flow in pipe. A typical value for pipe flow during tank truck loading is of the order 1 μA (Figure 5-3.1.1).

Various theoretical and empirical models have been derived expressing either charge density or charging current in terms of flow characteristics such as pipe diameter d (m) and flow velocity v (m/s). Liquid dielectric and physical properties appear in more complex models. The application of theoretical models is often limited by the nonavailability or inaccuracy of parameters needed to solve the equations. Empirical models are adequate in most cases. For turbulent flow of nonconductive liquid through a given pipe under conditions where the residence time is long compared with the relaxation time, it is found that the volumetric charge density Q_V attains a steady-state value which is directly proportional to flow velocity

$$Q_V = \alpha v \qquad (5\text{-}3.1.1a)$$

Where this equation is applied to different nonconductive liquids in different pipes, the polarity of the generated charge may change unpredictably and the proportionality constant α may vary over about an order of magnitude depending on conditions. The charging current I_C is the product of

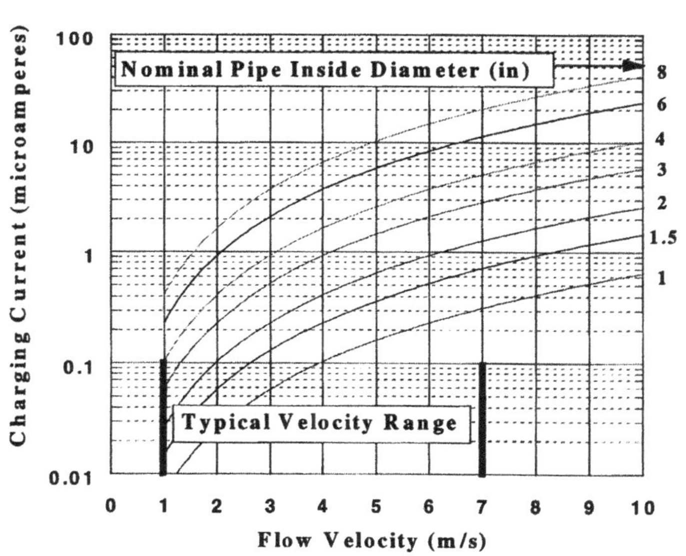

FIGURE 5-3.1.1 Predicted charging current for long pipe flow [Eq. (5-3.1.3)].

charge density (Q_V) and volume flow rate ($v\pi d^2/4$). If we assume that α might be a function of pipe diameter

$$I_C = \beta v^x d^y \tag{5-3.1.1b}$$

where β, x, and y are constants.

Various proposed values for the constants can be found in the literature [8]. Despite double-layer model predictions [148,149] that exponents x and y are both unity, and a dimensional analysis model [204] giving x as 1.88 and y as 0.88, test work on a practical scale [202,203] has indicated that both exponents are approximately equal to 2. This implies that α is roughly independent of pipe diameter and that the ratio $\alpha/\beta \cong 4/\pi \cong 1$.

The charging current I_C is therefore roughly proportional to the square of (vd), the velocity–diameter product. An important outcome is that the velocity–diameter product can be used to characterize charging current in pipe flow and as a basis for setting flow limits when filling tanks (5-4).

For Q_V in C/m^3, constant α has been reported in the range $(4.77-31.8) \times 10^{-6}$ Csm^{-4}. For I_C in Amperes, constant β has been reported in the range $(3.75-25) \times 10^{-6}$ Csm^{-4}; this follows directly from the relation $\alpha/\beta = 4/\pi$. The low values corresponded to turbulent flow through long, smooth pipe while the high values corresponded to turbulent flow through spiral-wound composite hose [8]. An order of magnitude value for both α and β is therefore 1×10^{-5} Csm^{-4}. As shown in [55], whatever assumptions are made concerning the charging current at the pipe wall, steady-state in turbulent flow is exponentially approached according to the modifying term $\{1 - \exp[-t/\tau]\}$, where the residence time t can be written as the length-to-velocity ratio L/v. This is reflected in Eqs. (5-3.1.2) and (5-3.1.3).

Assuming that the x and y exponents in Eq. (5-3.1.1.b) are indeed 2, it follows that to within a factor of about ±3 the charge density and charging current in pipe exponentially approach their long pipe values according to

$$Q_V\,(\mu\text{C/m}^3) \cong 10v\{1 - \exp[-L/(v\tau)]\} \tag{5-3.1.2}$$

$$I_C\,(\mu A) \cong 10\,(vd)^2\,\{1 - \exp[-L/(v\tau)]\} \tag{5-3.1.3}$$

where velocity v is given in m/s and the pipe inside diameter d and length L are given in meters. "Long" pipe has a liquid residence time (L/v) that is large compared with liquid relaxation time (τ), so that the exponential correction term is small compared with other errors and can be neglected. Figure 5-3.1.1 shows predicted "long" pipe charging currents for a series of pipe diameters, conveniently expressed in inches.

As discussed in 5-4.2 and 5-4.3, three different maximum velocity–diameter (vd) products can be found in current recommended practices. Values of 0.38 (this book) and 0.50 m^2/s (API 2003) have been recommended for road tanker filling. Values of 0.50 (this book) and 0.80 m^2/s (API 2003) have been recommended for rail car filling. From Eq. (5-3.1.3) these vd products (0.38, 0.50, and 0.80 m^2/s) yield predicted "long" pipe charging currents of 1.4, 2.5, and 6.4 μA, respectively. The charging current approximately doubles between these pairs of vd products. The significance of this with respect to faster loading times versus risk tolerance is discussed in 5-4.

An important practical question is, "what is the representative pipe diameter in loading circuits comprising different sizes of pipe?" This has a large effect on the values calculated for velocity and velocity-diameter product. As an example, static ignition of ester mist in a rail car (5-1.3.1) involved 1450 gpm through a 6-in. pipe (v = 5 m/s and vd = 0.76 m^2/s) followed by a short 4-in. dip pipe assembly (v = 11 m/s and vd = 1.15 m^2/s). Were nonconductive liquid flow rate restrictions applied to the semiconductive ester (time constant ~0.01 s) involved in this fire, the flow rate based on the 4-in. pipe would be unacceptably large based either on a 7 m/s maximum velocity or a 0.80 m^2/s maximum vd product. However, based on the 6-in. pipe upstream the flow velocity is less than 7 m/s and also meets API's $vd \leq 0.80$ m^2/s criterion.

The velocity–diameter product should usually be assessed for the pipe entering or terminating at the tank, neglecting restrictions such as tees or cruciform outlets on dip pipes. However, in top loading the pipe is often necked down at the rack to accept a smaller diameter hose followed by the dip pipe. The total length of smaller diameter hose plus pipe is only a few meters, which for liquids having time constants of several seconds does not constitute "long" pipe using Eq. (5-3.1.3). BS 5958 [2] assumes that only nonconductive liquids having time constants of about 1 s or more can accumulate hazardous charge densities in road and rail tankers. It therefore allows vd to be assessed on the basis of larger diameter upstream pipe in cases where the pipe entering the tank is relatively short and presumably unable to generate its correspondingly greater "long" pipe charging current. BS 5958 allows credit to be taken provided the smaller pipe is less than 10 m long and only one pipe size smaller than the larger pipe upstream (diameter not less than 67% that of the larger pipe).

In reality the charging current increases in the smaller diameter dip pipe. Also, small diameter spiral-wound hose can in some cases greatly increase charging current. This effect is presumed due to increased turbulence at the inner wall [8]. For overhead filling of road and rail tankers the maximum vd products recommended in this book are respectively 0.38 m^2/s and 0.50 m^2/s, which are smaller than the maximum recommended values given in

API 2003 and BS 5958. This took into account the possibility that credit might be taken for larger upstream pipe as described in BS 5958 and that operational errors such as partial dip pipe insertion might occur.

5-3.1.2. Pro-Static Agents

This concept was introduced to help explain the unexpected occurrence of "static" fires and explosions during fuel loading operations where there had been no apparent change in procedure. Should pro-static agents exist, their principal characteristic would be to increase the charging tendency without significantly increasing the conductivity. The charging current as governed, for example, by constant β in Eq. (5-3.1.1b), would be significantly increased while the rate of charge dissipation in a downstream tank, governed by conductivity, would remain unaffected. Hence, the rate of charge accumulation and magnitude of surface potential in the tank would increase. An unrecognized and equally important effect of a pro-static agent is to promote negative as opposed to positive charging. Brush discharges are only likely to cause ignition if they originate from negatively charged liquid surfaces (2-6.2).

The charge transfer mechanism in hydrocarbon fuels may be affected by a wide variety of additives and contaminants (2-3.1). An investigation concluded that certain fuel additives (corrosion inhibitors, thermal stability and antioxidant additives) can exert a pro-static charging effect [151]. The fuel components most electrostatically active were sodium sulfonates, particularly those derived from petroleum. The activity was highly dependent on chemical structure although there was a trend for the highest activity to occur at the lowest molecular weight in a given homologous series. With respect to the sulfonate observations, low-sulfur fuels are usually electrostatically active; it appears that the removal of electrostatically active species is more than compensated for by the accompanying conductivity reduction.

Suspended water is known to greatly increase charging as discussed in 5-3.6 and 5-3.7. However, it has been shown that dissolved water can be a highly effective pro-static agent [151]. Under some test conditions, involving filtration charging, the charging tendency due to dissolved water increased by about one order of magnitude while the fuel conductivity actually decreased. In general, dissolved water had a variable effect on conductivity of different hydrocarbons; the conductivity could increase, decrease or remain about the same. Since water did not increase the charging tendency of silica-gel treated *n*-heptane it was concluded that water is not a pro-static agent per se but instead interacts with some minor constituent in the fuel that is removable by silica gel. The effect of additives on conductivity may be synergistic (5-2.6) and it appears that synergy might also exist between pro-static agents, for example between a minor constituent and water. The possible effects of pro-static agents should be considered if velocity–diameter products greater than the recommended values (5-4) are used.

5-3.2. Nonconductive Pipe and Linings

Nonconductive plastic surfaces affect the rates of both generation and loss of charge during flow through pipe. Although the charging current in plastic pipe has been found to decrease with increased flow time it should be assumed that the charging rate is similar to that in metal pipe. Also, if the liquid is not conductive the pipe walls may become strongly charged. For nonconductive and some semiconductive liquids, insulation by the pipe wall can result in charge accumulation which may eventually lead to electrical breakdown and "pinhole" punctures either of the plastic lining or (for all-plastic pipe) the entire wall thickness. This has created problems in flammable liquid (such as xylene) and nonflammable liquid (such as silicon tetrachloride) service. For all-plastic above-ground pipe, external electric fields created by the accumulated charge may also cause static discharges from the external wall to neighboring conductive objects. These effects are generally not observed for conductive liquids ($\kappa > 10^4$ pS/m). During draining of lined pipe, nonconductive liquids might give rise to internal discharges, for example at metal connections, if the lining has accumulated a large surface charge density. Such discharges might ignite flammable mixtures inside the pipe. The hazard may be mitigated by using a semiconductive lining, blowing down the line with nitrogen, or providing a wait time for excessive charge to relax from the lining. The wait time recommended in section 19.8.3 of Part 2 of BS 5958 [2] is the shorter of the two times

a. t_1 = 3 relaxation times of the pipe lining material = $(3\rho\varepsilon_r\varepsilon_0)$ where ρ = volume resistivity of lining material in Ω-m and ε_r = dielectric constant of lining material. Note that dielectric constant is dimensionless and does not have units of Ω-m as given in the 1991 edition of BS 5958.
b. $t_2 = [L^2\varepsilon_r\varepsilon_0/(dr\kappa)]$ where L = distance between grounded points in contact with the liquid at ends of lined pipe section (m), d = lining thickness (m), r = pipe radius (m) and κ = liquid conductivity (S/m). Note that the 1991 edition of BS 5958 incorrectly expresses the units of liquid conductivity in pS/m. The equation as given in BS 5958 therefore yields results that are one trillion (10^{12}) times too small.

Both of these wait times can be impractically long. Using (a) the wait time is 2.2 hrs for a lining having a resistivity of 10^{14} Ω-m and dielectric constant of 3. Using the corrected equation (b), a 2-mm-thick lining of the same material in a 10-cm-radius pipe, 10 m between grounded points, being drained of toluene having a conductivity of 1 pS/m, has an impractical wait time of 154 days. The wait time given by (b) only achieves reasonable values for conductive liquids, in which case there should be no static accumulation problem in the first place. BS 5958 provides no rationale or literature reference for equation (b), which is evidently useless with or without correction.

Pipe with high resistivity lining that contains semiconductive or nonconductive flammable liquids should be blown down with nitrogen rather than air. To avoid pinhole damage, the flow rate during blow-down should be no higher than normal liquid flow rate. Also, the possible hazards created in downstream tanks by charged, two-phase flow should be considered (5-2.5.4).

5-3.2.1. All-Plastic Pipe

Owing to symmetry about the axis, a long plastic pipe whose walls are uniformly charged contains zero internal field, because the field contributions from diametrically opposite charges exactly cancel one another. Internal fields only exist where the field symmetry is perturbed near the pipe ends, or by conductive or charged objects external to the pipe. When a plastic pipe contains charged liquids or powders an internal field will be created, but in most cases the internal field is small owing to the small pipe diameter. Plastic pipes may create large external fields due to either internal or external charges. Because the internal field is small, very large internal charge densities can be accumulated without breakdown taking place; because the field is exerted radially through the pipe wall, electrical breakdown can only occur through the pipe wall itself. Since breakdown voltages can exceed 100 kV for a few millimeters of plastic, very large energies can be stored in plastic pipe walls before a breakdown occurs (2-6.5.3). PBDs may be produced from plastic pipe following internal charging either via flow of powder or charged liquid (2-6.5.2). Owing to end effects, a certain minimum length-to-diameter ratio may be required for PBDs to occur and puncturing is least likely to occur close to the pipe ends (5-3.2.3).

All-plastic nonconductive pipe such as polyolefin is not recommended for handling nonconductive or semiconductive liquids except where it can be shown that the advantages outweigh any risks associated with external static ignition or leakage via pinholes, or where tests have demonstrated that the phenomena will not occur. Burying an all-plastic pipe prevents external ignition risk in the buried section but will not necessarily rule out leakage due to pinholes should large charge densities accumulate. PVC and especially fiber-reinforced polyester (FRP) pipe should be considered separately from polyolefin pipe owing to the considerably smaller resistivities typical of these materials. The characteristics of FRP are highly variable and over time both the resistivity and breakdown voltage typically decline due to the development of minute cracks; this behavior is pronounced when FRP is used outside and subjected to freezing weather.

Plastic-lined metal pipe poses no external ignition hazard and will normally provide containment in the event of pinhole puncturing of the lining. However, the tolerability of pinholes in the lining should be considered. For example, if the liquid is corrosive to the metal pipe, gradual loss of metal via the pinholes might lead to unacceptable product contamination and eventual loss of containment. Conversely, minor pinhole damage may be acceptable if the lining is intended only to minimize product discoloration caused by rust and scale.

5-3.2.2. Mitigation for Plastic Pipe Hazards

Where nonconductive and semiconductive liquids must be transferred through plastic piping systems, mitigating strategies include

- Reduce rate of charging such as by decreasing flow velocity.
- Eliminate or relocate microfilters farther upstream.
- Reduce wall resistivity, possibly to less than 10^8 Ω-m.
- Increase breakdown voltage of wall by increasing thickness and/or changing material of construction.
- Incorporate external grounded conductive layer on pipe.

Combinations of these strategies might be considered. For example, in many cases the presence of an external conductive layer on a plastic pipe will not by itself eliminate puncturing of the internal plastic wall, and if the layer does not provide containment it will not prevent external leakage.

5-3.2.3. Plastic Tubing and Small Bore Hose

In 1959 a number of failures occurred in small bore hose used for aircraft fueling, which developed pinhole leaks caused by electrostatic puncturing. All of the hose was extruded Teflon tubing with braided stainless steel reinforcing wire [164]. Failures were found on operating aircraft and it was noted that the condition might go unnoticed for extended periods and then be misdiagnosed. All of the failures occurred in hose more than 18 in. long, no punctures occurred within 2 in. of the end fittings, times-to-failure varied from a few minutes to 2000 hours, multiple punctures could occur, and the punctures appeared like fractures when viewed under the microscope. It was estimated that voltages of about 50 kV were responsible for the electrical breakdowns. Following an extensive investigation, conductive Teflon was developed having the proper quantity and distribution of carbon black to dissipate the static while not adversely affecting other material properties. Later work [205] concluded "considering pipes or hoses through which a highly charged petroleum liquid may flow, an electrical resistance from each and any portion of the interior surface of the pipe to ground not exceeding about 10^7 Ω is adequate to limit pipe potentials resulting from static electricity to less than 1000 V."

Flexible tubing for high pressure service, equipped with stainless steel overbraid plus tube adapter end connections, is commonly available with a carbon black-loaded PTFE core tube to dissipate static. Numerous other designs of conductive and antistatic tubing are available for low pressure applications. The utility of conductive tubing in preventing fires during transfer of aromatic hydrocarbon liquids is described in [165].

5-3.3. Flexible Hoses

These are available in all-metal, lined metal, reinforced rubber/plastic, or composite ply types. Nonconductive hoses should not be used in flammable liquid service or in flammable vapor atmospheres. Also, if used immediately downstream of filters in nonconductive or semiconductive liquid service, hoses should be flexible metal, semiconductive or thick-walled conductive types. In some cases semiconductive linings might be necessary to prevent charge accumulation and pinhole puncture damage to conductive hoses. Conductive hoses should be electrically continuous via the embedded metal bonding element; the latter should be properly bonded to the end connectors and the continuity should be periodically checked. As shown schematically in Figure 5-3.5, the type of hose can affect static generation and there is evidence [8] that internal spiral-wound hoses of small diameter (1.5–2.0 in.) generate more static than smooth bore hoses, presumably due to greater turbulence at the wall. For vacuum truck operations, see 5-4.6.

5-3.3.1. Bonding and Grounding of Hoses

The resistance to ground from any point along a flexible metal conductive hose should be 10 Ω or less except where insulating flanges are employed to avoid stray current arcs.

Conductive hoses containing a continuous wire or braid bonding element should have a resistance ≤ 1 kΩ per meter of hose length and the resistance to ground from any metal connector should be less than 1 MΩ.

The resistance to ground through semiconductive hoses, whose current-limiting design eliminates a low resistance bonding element, and through insulating flanges, should be between 10^3 and 10^5 Ω per meter of hose length (or intermediate range as specified by a manufacturer) based on measurement between the end connectors; the total resistance to ground from a metallic hose connector should not exceed 1 MΩ.

A resistance to ground less than 1 MΩ will prevent static accumulation in practical cases. However, if periodic testing reveals a significant increase in the "as installed" value the hose and/or insulating flange should be evaluated for replacement. This is because the increased resistance could indicate corrosion or other damage that could lead to sudden loss of continuity. Where conductive hoses have double spirals (one for bonding and the other for mechanical strength) continuity between the end connectors need only imply continuity of one spiral. A fire was reported during draining of toluene from a road tanker through such a hose and after the event it was found that the inner spiral was not only broken but was not designed to be bonded to the end connectors. Two post-loading toluene fires occurred with a similar hose as the disconnected hoses were being handled by operators [166]. For nonconductive liquid handling especially, hoses should be selected with care. An option is to use a hose with a semiconductive or conductive lining so that a broken inner spiral cannot become isolated from ground and form a spark gap, and static rapidly drains from the hose after use. In any case the inner spiral should be designed so that it is firmly bonded to both end connectors. It is especially important to ensure continuity with end connectors (or nozzles) where a hose is used in a flammable atmosphere. In general it is safer to use a properly designed fixed filling system such as a dip pipe arrangement than to use a hose in tank filling operations.

5-3.3.2. Utility Hoses

These do not normally create a static hazard. However, where used in flammable vapor atmospheres such as inside tanks, these should be conductive or semiconductive and in particular it should be ensured that metal connectors and nozzles are grounded. See 5-3.3.1 for grounding requirements. Ungrounded hose connectors on nonconductive hose may become charged by a variety of mechanisms, such as by inserting a nitrogen hose into a tank containing charged liquid or mist. Although clean, dry gases do not generate charge, a nonconductive hose may become highly charged by the flow of steam. Typically, special conductive hoses are reserved for steam service.

5-3.4. Dip Pipes

See "Splash Filling" in 5-2.3.1. Dip pipes should be metallic, grounded, and permanently fixed where possible. Explosions have occurred where plastic dip pipes were used, both during filling and emptying operations. The vessels involved ranged in size from drums [8] to barge tanks [172]. The purpose of the dip pipe is either to load a vessel with minimal disruption of bottom sediment and mist production associated with splashing, or to empty a vessel. The dip pipe should extend close to the bottom of the vessel without touching it, and might be equipped with either a 45 degree cut or a tee to divert flow horizontally near the base of the vessel being filled. The design should prevent upward spraying during the initial stages of filling and this is typically achieved by using a "slow start" such that inlet velocity is held at less than 1 m/s until the dip pipe outlet is covered by at least 6 in. or two pipe diameters of liquid, whichever is greater. To prevent siphoning, a dip pipe may in some cases need to be equipped with an appropriate vacuum breaker such as breather holes.

5-3.4.1. Thrust Neutralizers

Sideways ejection of liquid at the foot of a dip pipe, via tees or cruciform outlets, can be used to reduce upward thrust forces in addition to the benefits noted in 5-3.4. However, it is important that the dip pipe not be partly inserted in a tank since this can exacerbate the static hazards. Sideways ejection of liquid via a right angle turn increases shear forces and static generation rate (5-1.3.1). If the pipe outlet is not fully submerged this will be accompanied by splashing and the production of mist or foam, which may result in a flammable atmosphere even below the flash point of the liquid being loaded. Also, the blunt end of a partly inserted dip pipe suspended above a charged liquid surface can be most effective in generating brush discharges.

5-3.5. Filters and Relaxation Tanks

Microfilters have a nominal pore size ranging from less than 1 μm up to 150 μm. Owing to the high solid–liquid contact area, large charging currents are generated in nonconductive and many semiconductive liquids. The charging current increases as pore size decreases and is frequently about 2 orders of magnitude greater than the charging current developed in the pipe upstream of the microfilter (Figure 5-3.5). Charge densities in nonviscous oils such as kerosene can attain 2000–3000 $\mu C/m^3$ with little dependence on flow rate [8], although charging current is at least proportional to flow rate

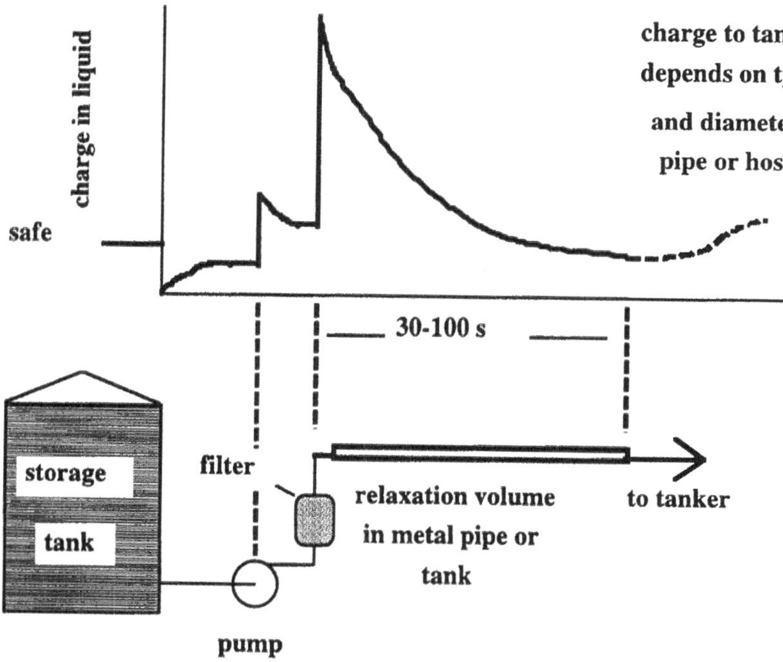

FIGURE 5-3.5. Schematic charge density variation in tanker loading circuit. (Adapted from API RP 2003.)

(A-5-3.5). Exceptionally large charge densities, up to about 5000 $\mu C/m^3$, may be generated by viscous nonconductive liquids (5-2.5.4).

To avoid high charging currents entering a receiving tank the filter must be placed sufficiently far upstream so that the charge decays to the ordinary value generated by pipe flow alone. Provision of three relaxation times, or a default 30–100 s, reduces charge by about 95% (5-2.4). Hence where the liquid conductivity is known, microfilters should be placed a minimum of 3 relaxation times upstream of a receiving vessel. Common industry practice is to provide 30–100 s residence time in conductive pipe or hose downstream of a microfilter in nonconductive liquid service where the conductivity is undefined. The higher value is theoretically desirable but might be impractical to achieve simply by flow rate reduction and/or filter relocation. Where it is not practical to supply three relaxation times or a default 30–100 s residence time in conductive pipe or hose, a "relaxation tank" may be used. This comprises a small, unpacked tank or section of large pipe providing increased residence time, in which the inlet and outlet pipes are oriented to avoid straight-through flow and the tank is designed to normally operate full of liquid. Packing the relaxation tank defeats its purpose (2-3). In the case of viscous nonconductive liquids, charge relaxation is too slow for these measures to be effective. Other mitigation such as inerting should be considered where a flammable atmosphere is present in the receiving tank (5-2.5.4).

Care should be taken when performing maintenance on microfilter elements wet with flammable liquid since sparks from ungrounded people or element parts during disassembly or element disposal have led to numerous fires. Where plastic bags are used for disposal they should be conductive or antistatic. In cases where spontaneous combustion might occur, for example where numerous filter elements containing combustible liquids are collected together, bags should be sealed after first (if compatible) wetting the contents with water.

5-3.5.1. Strainers

Mesh strainers finer than 100 mesh/inch (<150 μm) should be treated as microfilters. Coarser strainers up to 50 mesh/inch (300 μm) may generate significant static when fouled with accumulated debris, so should be treated as microfilters except in cases where fouling is not expected or may be rapidly identified by either periodic inspection or monitored pressure drop. "Clean" strainers should nevertheless be placed as far upstream as practical for nonconductive liquid service. A theoretical model for the charging process in strainers (screens) is given in [119–120]. Viscous nonconductive liquids (5-2.5.4) may produce unusually high charging currents in strainers.

5-3.5.2. End-of-Line "Polishing" Filters

These filters are sometimes used in cases where an upstream filter fails to provide adequate quality control, typically because of scale in the line between the filter location and the tank. Often such a filter is installed on the end of a hose and the filter element might be directly exposed to vapor in the tank. During cold weather, static discharges were observed on such a filter during transfer of semiconductive *n*-butyl acetate having a conductivity of about 2000 pS/m. Fortunately the liquid was well below its flash point. Unhoused polishing filters should not be used in flammable liquid service. For conductive and semiconductive liquids, the provision for three relaxation times (5-3.5) is simple to achieve with a grounded metal lance downstream of the filter. However, for nonconductive flammable liquid service a polishing filter should be used only in inerted atmospheres.

5-3.6. Suspended Material

Aeration of liquids not only increases charging due to the large liquid–bubble interfacial area but can also produce hazardous conditions in downstream equipment. Even if a downstream tank is inerted, the incoming aerated liquid may produce an ignitable froth (5-1.3.1). Immiscible or poorly soluble liquids may disperse as droplets or as an emulsion. Where a nonconductive liquid contains a dispersed phase such as water in oil, the continuous phase determines the charge relaxation behavior while the charging behavior is related to the electrical double layer formed at the interface between the two phases. The total interfacial area of the dispersed material can be very high compared with the boundary contact area where charging would otherwise occur. Hence, charge generation is generally greater for such suspensions than for a single phase liquid. Double layer shear can occur during flow, pumping, filtration or settling of suspended liquids or solids to the base of a tank. For water in oil, the effect of the second phase is appreciable above about 0.5 wt%. Where a dispersed phase is expected, additional precautions such as the reduction of flow velocity to about 1 m/s, or inerting, should be considered. Where lines contain low points that might collect water, the flow velocity should not be maintained at less than 1 m/s for extended periods. This is because a large concentration of water previously collected in the low points could be flushed out when the flow rate is increased. For mixing operations see 5-4.4.

5-3.7. Valves and Other Line Restrictions

Items such as orifice plates, valves, elbows and tees increase turbulence and may result in increased charge density downstream. Suspended material such as water (5-3.6) has been found to greatly increase this effect; for example 1–5 wt% water in toluene produced 580–620 μC/m downstream of a partly open 4-in. ball valve [20], which was about 20 times the charge density observed for pure toluene flow under the same conditions. Where a nonconductive liquid contains upward of 0.1% by volume of free water and this type of flow restriction exists immediately upstream of a receiving tank, flow velocity should be restricted to about 1 m/s or less [20]. This compares with the more general recommendation (5-3.6) to apply this restriction above 0.5 wt% free water in unrestricted pipe flow. Smaller charge densities are generally produced during unrestricted, straight pipe flow. Another reported effect of ball valve electrification is sparking from valve handles insulated by the packing; special ball valves have been designed which provide a resistance less than 10 Ω between ball, spindle and outer casing [19]. If electrical continuity with the valve body is not found with the handle in all possible positions, the handle should be separately bonded.

5-4. Filling Criteria for Tank Operations

Filling criteria for grounded metal tanks are intended to restrict the maximum liquid surface potential to less than the threshold value at which incendive brush discharges can occur to a grounded electrode above the liquid (A-5-4). As a tank is filled with charged liquid the surface potential increases and usually passes through a maximum value at the center of the free liquid surface when the tank is 30–50% filled. It is assumed that the charged liquid in the tank has a small conductivity and is well mixed so that the charge density is approximately uniform (5-2.4).

The relationship for liquid surface potential Φ during tank filling is of the form [22]

$$\Phi = kQ_V f\,(\text{tank dimensions and liquid depth}) \qquad (5\text{-}4.1)$$

where k is a constant and Q_V is the charge density in the tank, given as a time-dependent variable by

$$Q_V(t) = (I_C\tau / V_t)\{1 - \exp(-t/\tau)\} \qquad (5\text{-}2.4.4)$$

The liquid surface potential is proportional to the charge density in the tank (Q_V), which in turn is roughly proportional to the charging current (I_C). The charging current is roughly proportional to the square of vd, the velocity–diameter product (5-3.1.1). The maximum liquid surface potential therefore varies approximately with the square of the velocity–diameter product of the liquid in the filling pipe. A relatively small increase in velocity can therefore result in a magnified increase in ignition probability as reflected by an increased surface potential. Neglecting any slow start, if a 7000-gallon road tanker is filled through 4-in. pipe, a vd of 0.38 m^2/s corresponds to a filling time of about 14.6 min and a vd of 0.50 m^2/s corresponds to a filling time of 11.1 min. In order to save this 3.5 min of filling time, the vd product has been increased by a factor of 1.32 and the surface potential in the tank has been increased approximately by the square of this number, or 1.73. This simplified analysis shows that there can be a large assumption of risk for only small benefit in reduced filling time.

Since velocity varies with the inverse square of pipe diameter d, an important consideration is the selection of pipe diameter. For any given velocity-diameter product, larger pipe diameters allow larger flow rates. Since occasional static ignitions in road tankers may occur at $vd = 0.38\ m^2/s$, smaller values might be considered for nonconductive liquid transfer depending on risk tolerance.

Table 5-4 gives solutions for various tank filling criteria discussed in this chapter in which either a maximum velocity or a maximum velocity–diame-

***TABLE** 5-4. Solutions (Gallons per Minute) for Various Tank Filling Criteria*

Diameter (in.)	v = 1 m/s	v = 7 m/s	vd = 0.38 m²/s	vd = 0.50 m²/s	vd = 0.80 m²/s
1	8	56	(120)	(158)	(253)
1.5	18	126	(180)	(237)	(379)
2	32	225	(240)	(316)	(506)
2.5	50	351	300	(395)	(632)
3	72	506	360	474	(758)
4	128	899	480	632	(1011)
6	289	2023	720	948	1517
8	514	3596	961	1264	2022

ter product is specified. The solutions are in US gallons per minute for a series of pipe diameters in inches. The 7 m/s maximum velocity criterion takes precedence over velocity–diameter products; the latter are parenthesized where the corresponding velocity exceeds 7 m/s. To avoid this problem when applying a velocity–diameter product of ≤ 0.8 m^2/s to rail car filling, API 2003 now tentatively allows flow velocities up to 10 m/s.

An important consideration for safe tank operation is that conductors capable of acting as electrodes are not suspended above a highly charged liquid surface, as demonstrated in the 1950s by full-scale tank ignition tests [55]. Thus, fill pipes should extend close to the floor of the tank so that they will be submerged before large surface potentials are generated (5-3.4). Electrically isolated conducting objects associated with floating debris, gauge tape flotation devices and suspended sampling containers should be avoided. Slow velocity starts at less than 3 ft/s or 1 m/s are often employed to reduce charge density until the fill pipe is submerged to a prescribed depth varying from about 6 in. for fixed tanks and 2 ft. for marine vessel tanks, allowing for sloshing. Additional benefits of a slow start are reduction of splashing, with its attendant froth and liquid mist formation, reduction of turbulence, which might create suspensions of water or other heavy material from the bottom of the tank, and avoidance of excessive charge accumulation at the liquid surface. It is important to note that submergence of the dip pipe does not by itself rule out incendive brush discharges at high charging currents. Several studies have demonstrated either large static discharges or gas ignitions at submerged electrodes and filling pipes [35, 71, 141].

Another important consideration is the allowance of a "wait time" before introducing objects that could act as electrodes, such as sample containers and gauge rods. The maximum liquid surface potential typically occurs when the tank is 30–50% filled and is proportional to the charge density in the tank. As flow to the tank is started or stopped the charge density rapidly rises to a geometry-dependent maximum value or falls to zero, as governed by the dimensionless group t/τ in Eqs. (5-2.4.4) and (2-3.7). As discussed in [195] the surface potential on a liquid with a conductivity of 15 pS/m rises to its maximum value or falls close to zero in only a few seconds. An important consequence is that discharged liquid already in the tank does not safely "dilute" charged liquid flowing to the tank. If flow is restarted after being stopped, the surface potential quickly regains or exceeds the value that previously existed. For example, multistage loading of road tankers can result in transient surface potentials that are greater than those produced by continuous filling; a number of loading fires have occurred under such conditions [195]. The worst case relaxation time is in the range 30–100 s (5-2.4.1), so wait times of 5–10 min. are adequate for relatively small tanks such as road and rail tankers. In the case of large storage tanks and ships' tanks a 30-min wait time is recommended to allow for complicating effects such as settling potential (5-4.1.2).

5-4.1. Storage Tanks

The maximum liquid surface potential attained in tanks depends not only on the liquid charge density, tank volume and filling rate, but also on the tank dimensions [69,191]. For commercial tanks of equal volume, the maximum potential is greater in tanks having smaller cross-sectional areas since liquid depth is built up faster relative to the rate of charge relaxation. Smaller potentials are therefore generated in a barge tank than in a vertical cylindrical tank of equal volume. The competitive rates of charge input and decay vary with tank dimensions and liquid properties while the loading rates and pipe diameters used in practice also vary with tank volume. Hence, no single set of filling recommendations, such as a maximum flow rate for a specified inlet pipe diameter, can efficiently address all sizes and shapes of tank. The liquid surface potential is proportional to the charge density in the tank, which under well-stirred conditions varies with time according to

$$Q_V(t) = (I_C\tau/V_t)\{1 - \exp(-t/\tau)\} \qquad (5\text{-}2.4.4)$$

Liquid relaxation times are normally less than 30–100 s while tank filling times generally exceed 1000 s. The exponential term in Eq. (5-2.4.4) is therefore negligible by the time most tanks are 30–50% filled. Hence

$$Q_V(t) \cong I_C\tau/V_t \tag{5-4.1.1}$$

or

$$Q_V(t) \cong I_C\tau/(Ft) \tag{5-4.1.2}$$

where F is the volume flow rate and t the time to accumulate volume V_t. From Eq. (5-2.2) the charging current I_C can be replaced by $FQ_V(0)$ where $Q_V(0)$ is the charge density entering the tank. Hence

$$Q_V(t)/Q_V(0) \cong \tau/t \tag{5-4.1.3}$$

Equation (5-4.1.3) shows that the fraction of the incoming charge density remaining in the liquid decreases during filling and achieves very small values where $t >> \tau$, such as in large tanks that are 30–50% filled. This decreasing charge density offsets the larger potentials attained in larger tanks due strictly to their greater cross-sectional areas. Another mitigating factor in large tanks is that the incoming charge density is proportional to flow velocity [Eq. (5-3.1.2)] and therefore cannot be indefinitely increased; apart from practical limitations, it has for many years been common practice to limit flow velocity to less than 7 m/s. Worst case conditions in terms of tank size and charge density should develop at some intermediate tank volume; depending on tank shape and other factors this might be in the range 10^3–10^4 gallons.

Large storage tanks usually present less hazard than those of intermediate size such as road tankers, since it becomes impractical to fill them sufficiently fast to offset charge relaxation as liquid level builds. Use of large-diameter pipe tends to offset charging caused by high flow rates (Table 5-4). It is sufficient to limit flow velocity to less than 7 m/s for tanks exceeding about 100 m^3 or 26,000 gallons; a typical volume for a rail car is 23,000 gallons. For large-diameter lines supplying large tanks this velocity is often limited by the pump capacity. Because pumping rates can be difficult to control procedurally, avoidance of excessive flow rates might need to be engineered into the system design.

5-4.1.1. Floating Roof Tanks

These tanks are designed to reduce evaporation and vapor emission. Two distinct designs are those with and those without an additional fixed roof. Of the former types, known as internal floating roof tanks, various provisions may be made for ventilating the space between the floating roof and the fixed roof. Fires may occur around the rim of a floating roof, and if adequate ventilation is not provided flammable vapor may accumulate between the floating roof and a fixed roof. Apart from lightning, static electricity is an issue only when a grounded, conductive floating roof is not afloat (that is, when low liquid level results in the roof settling on its supports) or personnel need to be on the upper side of the roof. When the roof is not afloat, liquids should be loaded at 1 m/s or less, after which the velocity can be increased to the normal design rate. Excessive flow rates and conditions such as aeration may cause a floating roof to tilt or even sink. In such an event, it may be necessary to apply foam to the liquid surface to suppress evaporation and reduce fire risk. If this is done, the foam should be applied at, and tangential to, the wall so that it spreads from the grounded wall toward the tank center without creating isolated patches. Fires have been sparked by isolated patches of conductive foam due to residual charge in the underlying liquid. The floating roof engages the tank wall via special "shoes" which in practice may fail to provide proper grounding. A separate grounding cable is sometimes used to bond the roof internally to its internal supports. When receiving flammable liquid such as gasoline into an empty internal floating roof tank it should be assumed that a flammable vapor space may persist below the fixed roof for 12–18 h after the floating roof has begun to float [156]. The area on top of the floating roof is considered a confined space and the atmosphere must be tested for flammable vapor prior to and often during any activity involving personnel on the roof.

5-4.1.2. Settling Potential

Gravitational separation of nonmiscible phases such as oil–water, solid–liquid or liquid–air mixtures results in charge separation. In the case of solid particles in liquids the phenomenon is sometimes known as the "Dorn effect." The mechanism involves formation of an electrical double layer around the dispersed particles comprising an inner layer of charge which is more tightly bound than an outer "diffuse" layer. Movement of the particles relative to the continuous phase shears the inner from the outer charge layer, causing a net charge to be carried by the dispersed phase [55,148]. The best known example is transfer of water-oil mixtures into large tanks, causing the generation of a "settling potential" as the water droplets settle out. This is an unlikely ignition mechanism for most chemical operations, especially if measures are taken to avoid suspended material (5-3.6 and 5-3.7). Settling hazards are increased by decreased conductivity of the continuous phase, increased concentration of suspended material, increased density difference between the phases (causing more rapid separation) and increased tank capacity, at least 100 m^3 probably being required. The situation is less severe than in liquid–solid mixers and other equipment that creates rapid rates of surface shear between the disparate phases (5-4.5). An analogous effect may occur due to gas bubbles rising in large tanks, for example if lines are blown down with air (5-2.5.4).

5-4.1.3. Tanks with Nonconductive Linings

For tanks having liquid recirculation see also 5-4.1.4. For reactors and other vessels see 5-4.4.1. For cleaning operations see 5-6.1.2. In some cases it is possible to specify a conductive lining material such as an applied coating (5-4.4.1). The hazard of filling tanks having nonconductive tank linings depends on various factors including lining thickness and resistivity, tank volume, liquid conductivity, operation involved and the ignition sensitivity of the tank atmosphere. Owing to the large number of variables the precise situation needs to be evaluated.

The simplest and most usual case is where the liquid is sufficiently conductive to have a constant potential. If it is assumed that the tank is splash-filled with no grounded internals, the incoming charge accumulates at the wall and generates a potential governed by the capacitance of the lining. If the lining is a perfect insulator, the equilibrium voltage across it can readily be found as illustrated for lined drums in 5-8.2. In most cases however, charge will be lost through the lining according to Ohm's law. This can be analyzed using Eq. (4-1.3) which treats the liquid as a poorly grounded capacitor, or parallel RC circuit, capable of producing sparks

$$R_L < I_C^{-1}\,(2W/C)^{0.5} \tag{4-1.3}$$

Leakage resistance

$$R_L = d\rho/A \tag{2-3.4}$$

Tank capacitance

$$C = A\varepsilon_r\varepsilon_0/d \tag{2-3.5}$$

Hence

$$R_L < I_C^{-1}\{2Wd/(A\varepsilon_r\varepsilon_0)\}^{0.5} \tag{5-4.1.4}$$

As the tank fills the capacitance increases while the leakage resistance decreases in proportion to the wetted area. The conservative case is to consider only the floor area. Assume that the charging current $I_C \leq 10\ \mu A$ and that the gas involved is hydrogen, whose LMIE $W = 0.016$ mJ. For lining thickness $d = 2$ mm, dielectric constant $\varepsilon_r = 4$, and floor area $A = 10\ m^2$, the resistance R_L should be less than 1.3 MΩ. This corresponds to a lining resistivity ρ less than 6.5×10^9 Ω-m, which might for example be achieved with glass reinforced polyester. Similar calculations for a less sensitive gas such as methane ($W = 0.21$ mJ) give respective values of 4.9 MΩ and 2.4×10^{10} Ω-m. From Ohm's law the maximum liquid potential is about 10 volts per megohm of resistance, allowing a roughly tenfold margin of safety relative to the minimum ignition voltages of high capacitance sparks given in A-4-1.3. Also, the assumed charging current should be conservative (5-3.1.1). A conservative guideline is that the floor resistance should be the order 1 MΩ or less corresponding to a resistivity of 10^{10} Ω-m or less. Equation (5-4.1.4) might be used to justify higher values, for example where a much thinner lining or smaller charging current are involved. If the tank is equipped with a compatible grounding system such as a grounded plate, the allowable lining resistance is much greater as described in 5-4.7.

If the liquid is nonconductive and therefore unable to generate sparks from its surface, the criterion given by Eq. (5-4.1.4) does not apply. Instead the lining should be selected to introduce negligible additional impediment to charge dissipation from the liquid, implying a lining time constant ($\tau = \varepsilon_r \varepsilon_0 \rho$) equal to or smaller than that of the liquid. Assuming the liquid has a time constant of 1 s, this criterion suggests that the resistivity of a lining with dielectric constant of 4 should be less than about 10^{10} Ω-m. No simple model is available for maximum potential in the tank, since the effect of a lining is superimposed on an already complex situation for the unlined tank (5-4). Painted coatings having a resistivity up to at least 10^{12} Ω-m should introduce negligible additional hazards. Typical thicknesses of such coatings are less than 7 mils (0.2 mm). Since charge accumulation on the lining is limited by its breakdown voltage, breakdown should occur at a small fraction of the >20 kV liquid surface potential at which brush discharges are observed (A-5-4).

The probability of a PBD on high resistivity linings should be negligible unless the incoming stream is directed against the tank wall and the lining meets the criteria given in 2-6.5. However, pinhole puncturing of the lining is predicted if the voltage across it exceeds its breakdown voltage V_b. Neglecting imperfections in the lining (5-4.4.1), an order-of-magnitude condition for avoiding pinholes is obtained from Ohm's law plus Eq. (2-3.4)

$$V_b > I_C d\rho/A \qquad (5\text{-}4.1.5)$$

or

$$E_b > I_C \rho/A \qquad (5\text{-}4.1.6)$$

Some typical values for breakdown strength E_b are tabulated in Appendix B. For example, a resistivity of $<10^{13}$ Ω-m might be required to avoid pinholes in a lining whose breakdown strength $E_b = 20$ kV/mm, assuming $I_C = 10\,\mu A$ and $A = 10\,m^2$. Note that this evaluation does not address ignition hazards.

5-4.1.4. Tanks with Recirculation

Liquid recirculation may produce combinations of adverse hazardous effects involving the accumulation of both static electricity and desorbed gases. The former effect is magnified if the tank contains an internal nonconductive lining or coating. These points are best illustrated by a case history. On December 4, 1998 in Texas, a dry crude ethylene dichloride (EDC) buffer tank exploded. The 18-month-old tank was 40 ft. tall by 60 ft. diameter and contained a 14-mil-thick polymer coating to prevent the corrosion that had been experienced in the old tank. EDC flowed from the reactor into the nitrogen-padded tank at 450 gpm through an uncoated 6 in. dip pipe terminating about 18 in. from the floor. EDC was pumped out through a side mounted 6-in. pipe, centered 16 in. above the floor. While flowing from the reactor to the tank the EDC experienced a drop in pressure from 20 to 0 psig, causing degassing. Degassing was probably exacerbated by a low (~5%) liquid level in the tank, causing intermittent splash filling and creating a vortex at the exposed outlet pipe. Evidently sufficient degassed oxygen accumulated over a period of several days to defeat the nitrogen padding and produce a flammable atmosphere. Subsequent investigation showed that oxygen derived from the reactor chlorine feed was unexpectedly soluble (~40 ppmw) in EDC. Ethylene co-reactant at an estimated 5 ppmw also dissolved in the EDC. Hence, ignition might have involved a local accumulation of degassed ethylene plus oxygen. A primary explosion in the base of the tank was followed by a secondary explosion as additional air entered the tank. Since EDC is a semiconductive liquid (~4000 pS/m) the accumulation of static was related to the combined effects of recirculation, low liquid level causing splash filling, and a nonconductive coating. The static discharge most likely occurred from charged liquid in the base of the tank to the end of the dip pipe. Alternatively a discharge might have occurred to an uncoated part of the outlet pipe, either from the liquid vortex or from a region of coated wall that was intermittently washed with charged liquid.

5-4.1.5. "Skim Layer" Hazards

In some cases a conductive liquid contains a small quantity of lighter, immiscible nonconductive liquid contaminant. An example is an inorganic acid containing a small quantity of oil which forms a nonconductive "skim layer" after transfer to a tank. To explain fires during the filling of unlined tanks it was theorized that during splash-filling the skim layer acts as the dielectric in a capacitor formed between the underlying acid and small, unstable pools of charged acid floating on the oil layer. The charge density of the acid is magnified if the inlet line is blown down with air at the end of the acid unloading cycle, as was typical for these operations in the 1950s; this plus a presumably thicker oil layer and greater vapor concentration may explain the timing of the ignitions. The proposed ignition mechanism was production of sparks during the collapse of unstable supported acid pools into the main grounded body of acid [175]. The source of a flammable atmosphere was typically light hydrocarbons derived from the oil. Hydrogen generated via acid reactions might explain ignitions in other cases. To avoid skim layer hazards, bottom filling should be considered.

5-4.2. *Road Tankers*

See 5-4. Road tankers have volumes less than 50 m^3. A typical volume is 7,000 gallons (26 m^3), corresponding to a cylinder roughly 5.4 ft. diameter by 42 ft. long. Single compartment tanks have a typical *L/D* ratio of about 7.8, although multicompartmented tanks have much smaller individual *L/D* ratios. More than 60 static ignition incidents were cited in a 1967 API survey, of which roughly 90% involved a static discharge within the truck compartment as a result of charge carried by the oil during loading [108]. An extensive theoretical evaluation of road tanker electrostatic hazards is given in [190–194]. The analysis is based on a model that assumes that liquids are sufficiently well-stirred to give uniform bulk charge density inside the tank and that a surface potential of at least 60 kV is needed for ignition in the presence of an ideal electrode (see A-5-4). Brush discharges are produced above 20–25 kV. Unless such discharges are seen as a tolerable risk the charge density thresholds should be 2–3 times smaller than those adopted in the analysis.

The tanker should be inspected for ungrounded components or foreign objects, and bonding components such as clips should be periodically examined. Bonding and grounding should be in place prior to starting operations; for example bonding of a dip pipe to the tank shell should be done before inserting the dip pipe in the tank. The use of alarmed or interlocked ground indicators can be used to ensure bonding is in place. Hatches should be closed on tank compartments except those that need to be open, and operators should be kept off trucks during loading [108]. However, even with proper bonding and grounding in place, excessive charge densities entering the truck may cause brush discharges from the liquid surface.

Microfilters should be grounded and placed a minimum distance upstream corresponding to three relaxation times of the liquid at maximum loading rate (Figure 5-3.5). This typically requires at least 30 s residence time in the line between filter and tanker for nonconductive liquids although the required residence time is negligible for conductive liquids in conductive pipe or hose. For semiconductive liquids having conductivities in the range 1000–5000 pS/m, static problems have been experienced in cases where nonconductive hoses or end-of-line polishing filters have been used (5-3.5.2); such problems are easily corrected since only about 0.3 s residence time in conductive pipe or hose is required to provide three relaxation times. For semiconductive liquids of lower conductivity (100–1000 pS/m) the required residence time for filter placement may need to be increased to 3 s. Avoidance of excessive flow rates should be ensured either procedurally or, preferably, by system design. For two-phase systems such as a hydrocarbon containing undissolved water see 5-3.6.

For road tankers with nonconductive coatings or linings the possibility of static accumulation on the inner surface causing pinhole damage or even a PBD should be considered (2-6.5, 5-4.1.3, and 5-4.4.1). However, no case history of a PBD is available for lined liquid tanks.

a. *Top Filling.* Splash filling with attendant production of charged mist and foam should be avoided using a dip pipe (5-3.4). To minimize splashing a slow start (less than 1 m/s) should be employed for nonconductive liquids until the fill pipe outlet is covered to a depth of either two pipe diameters or 6 in. of liquid, whichever is larger. A slow start should be considered for semiconductive liquids where partial insertion of the fill pipe cannot be ruled out, especially where a tee or thrust neutralizer is used (5-3.4.1). Transition from slow start to normal pumping rate might be achieved automatically using a special loading regulator tip, which shifts rate when submerged to a prescribed depth. The product of flow velocity and inside diameter of the filling pipe may then be increased to $vd \leq 0.38\ m^2/s$ for pure liquids such as toluene, whose vapor will normally be in the flammable range throughout filling. Equation (5-3.1.3) predicts that within a factor of about ±3, a vd product of $0.38\ m^2/s$ yields a charging current of $1.4\ \mu A$ in long pipe. This same velocity–diameter product should be used for switch loading operations except where the liquid being loaded has an unusually large relaxation time such as some viscous nonconductive liquids (5-2.5.4). Unless antistatic additives can be used to decrease the relaxation times of these liquids, switch loading should be done only under inerted conditions. In no case should the flow velocity exceed 23 ft./s (7 m/s). A wait period of at least 5 min should be observed before removing the dip pipe and no sampling or gauging should be performed for at least 10 min after cessation of flow. However, sampling via a sample well or "gauge well" can be done at any time.

Note 1. When loading volatile products such as gasoline, whose vapor concentration can be shown to rapidly exceed the upper flammable limit during tank filling, the velocity–diameter product may be increased to 0.50 m^2/s. This is consistent with API RP2003 [3]. Similarly, shorter wait periods of 1–2 min can be used.

Note 2. A fixed dip pipe avoids possible hazards of partial dip pipe insertion (5-1.3.1).

Note 3. Concerning the adequacy of recommended vd products, there have been no realistic studies using pure liquids at temperatures producing their most easily ignitable vapor concentrations, for example toluene at 26°C (5-1.4.1). As discussed in 5-4 and elsewhere, recommended tank filling criteria combine theoretical modeling with limited experimental studies, augmented by loss statistics available to those formulating the criteria. Static discharges are most likely to occur when tanks are 30–50% full, at which point vapor from petroleum distillates such as gasoline will normally be too rich to burn. When considering the statistics of tank truck fires, pure liquids should always be considered separately because switch loading is always less hazardous than loading a pure liquid at a temperature corresponding to the most easily ignitable vapor concentration (5-1.4.3).

b. *Bottom Connection Filling*. The bottom loading inlet "footvalve" should be designed with appropriate deflectors such as caps to direct incoming liquid sideways toward the tank walls rather than upward, which will create spraying and mist production. The practical effect of an upward-directed jet may be to increase the charge density at the liquid surface by directing unrelaxed liquid there and possibly encouraging static discharges. The recommended maximum velocity–diameter product is 0.38 m^2/s and, owing to the absence of the grounded dip pipe, the value adopted should theoretically be about 25% smaller than would be used for top filling the same container through a centrally located dip pipe.

c. *Combustible Liquids*. Bonding and grounding should be considered for Class II and Class III liquids loaded at racks even where no Class I flammable liquids are handled. If switch loading (5-1.4.3) is a possibility the vapor space of a tanker and connected tank may be flammable even if the liquid transferred has a high flash point. Also, unless measures are taken to prevent splashing and attendant production of foam and mist, a flammable atmosphere may be produced irrespective of flash point (5-1.3).

d. *Closed Connection Transfers.* A closed connection is where a hose is connected before start of flow and valved off before connection is broken. Hence there is normally no flammable atmosphere created external to the tanker or hose. Where vehicles are loaded or unloaded through closed connections a static hazard may exist if the tanker is drained of a nonconductive liquid, since as air is sucked in, a flammable vapor–air mixture may be created in a hose previously charged by liquid flow (5-3.3.1). Also, the potential for leaks and spills always exists for hose transfers. It is therefore recommended that the tanker be grounded and that nonconductive hose not be used for Class I liquids.

e. *Highway Transport.* As noted in API RP2003, road tankers normally create no static hazard during transport provided they are either full, compartmented or contain baffles. This minimizes sloshing of the liquid in the tanker due to changes in acceleration, which might result in charge accumulation. Nonconductive liquid fires in partly filled, unbaffled tankers have been reported, although instances are very rare. Unbaffled, single compartment tankers used for nonconductive liquids capable of generating flammable atmospheres should preferably employ tanks that are full or have small outage. While API RP2003 limits this restriction to intermediate vapor pressure products, consideration should be given to possible mist ignition and other factors that might reduce the assumed flash point of combustible liquids such as kerosene (5-1.1).

5-4.2.1. Effect of Road Tires

Large potentials up to about 100 kV can accumulate on moving trucks due to contact-separation of tires with the pavement (A-5-4.2.1). On stopping, charge drains rapidly through the tires to ground and potentials typically decrease to a nonhazardous level in about 10 s. There is negligible practical hazard from charging via road tires and for this reason the requirement for drag chains was deleted from NFPA 385 in 1953. This concern is entirely separate from charging during liquid transfer, which requires bonding and grounding.

5-4.3. Rail Cars

See 5-4.2. Rail cars have volumes of the order 100 m^3. A typical volume is 23,500 gallons (89 m^3), corresponding to a cylinder approximately 9 ft. diameter by 49 ft. long. Single compartmented tanks are typically unbaffled with *L/D* ratio about 5.4. Of the roughly 220,000 cars in North America, less than about 2% are compartmented; these mostly have two compartments although some have three (four is rare). The larger typical volume of single compartment rail cars (23,000 gallon) compared with road tankers (7000 gallon) allows greater filling rates to be used. A variety of velocity–diameter (vd) limits have been recommended; the largest current value being API's $vd \leq 0.80$ m^2/s for petroleum products [3]. In this book the recommended value is $vd \leq 0.50$ m^2/s for filling nonconductive flammable chemicals, especially in the case of compartmented cars. Semiconductive chemicals should also be subject to filling rate restrictions; at a minimum the flow velocity should not exceed 7 m/s. These restrictions reflect in part that fires involving fine chemicals typically have greater consequence than those involving commodity products such as fuels. If filling time is an important economic factor, a larger diameter loading line might be considered to allow faster loading rates without increasing the vd product. Also, the use of a fixed dip pipe should be considered so that splash filling via operational error cannot occur.

Bonding between the wheels and rails can be assumed to be inherently present by direct contact, but car components might be isolated from each other by nonconductive bearings and wear pads. Where required these components should be bonded together and grounded to the loading rack ground with an electrical continuity of less than 1MΩ. Additional measures may be required if stray current protection is provided [1,3]. If insulating flanges are present in the loading line or vapor recovery line, grounding on the rack side of the flange should be permanently installed and bonding to the car on the car side of the flange should be made before commencement of filling. If present, insulating flanges must not be jumpered or otherwise short circuited.

As in the case of road tankers there are no available case histories of PBDs in lined rail cars for liquids. However, owing to the large charge densities developed in some powder transfers the effect of linings may need serious consideration for rail cars in powder service (6-4.3).

5-4.4. Liquid Phase Mixers, Blenders, and Reactors

Where possible, agitators should be covered with sufficient depth of liquid to minimize splashing, or operated at a reduced speed until sufficient depth has been built up. In cases where hazardous static accumulation in the liquid cannot be avoided using the measures discussed here, ignition hazards can be mitigated by inerting plus avoidance of metal projections from the vessel head such as thermowells that could act as discharge electrodes above charged liquid surfaces.

Static accumulation may result from stirring, splash-blending or recirculation, disruption of the liquid interface by a jet from below, or loading practices covered previously under "Tanks." Where a conductive and nonconductive liquid are to be blended, the conductive liquid should be added first where possible, so that the resulting stirred blend is semiconductive or conductive. Recirculation loop reentry should be designed not to cause splashing or surface disruption, for example, by using subsurface jets that do not break the liquid surface. It can be assumed that liquid entering a tank has a greater charge density than liquid already in the tank, which has had time to relax its charge. Therefore, when introducing two or more nonconductive liquids to a blending tank, the less dense liquid should where possible be loaded first to avoid a surface layer comprising the lighter, more highly charged component. Splash recirculation of nonconductive liquids should normally be done only if the vessel is effectively inerted. Under normal conditions many operations involving gasoline blending are nonflammable since volatile fuel components result in mixtures above the UFL. Prior to establishment of vapor–liquid equilibrium a slow start procedure can be followed.

5-4.4.1. Vessels with Nonconductive Linings

The use of conductive linings and coatings should be considered where liquids are handled in potentially flammable atmospheres, especially where the liquids are semiconductive or nonconductive. Although conductive coatings are available these might be ruled out by noncompatibility. There are additional difficulties finding suitable conductive enamels for reactors. For vessels already in service with nonconductive coatings or linings, the possible risks due to static accumulation should be assessed and mitigation measures such as nitrogen inerting should be considered.

Static charge accumulation on linings may result in "pinhole" lining damage in equipment such as enamel or glass-lined reactors [157,159]. In addition to replacement costs and possible quality problems, a major hazard could result if iron is catalytic to the process, especially if subsequent reactor conditions such as low pH allow rapid enlargement of pinholes [157]. In 1981 a runaway reaction and ensuing major fire occurred via this mechanism during the production of vinyltrimethoxysilane. A contributing factor might have been pinhole formation caused by the stirring of poorly conducting mixtures, possibly during the hexane solvent addition step.

Since static discharges frequently occur at the liquid interface as liquid drains from the wetted wall, a vapor ignition hazard may also exist [157,159]. These discharges may be either brush types or PBDs (2-6.2 and 2-6.5). In 1997 it was concluded that "Manufacturers have so far not paid attention to requests by the operators of enamel apparatus for a coating which is completely or partially capable of dissipation" [159]. However, this situation is believed to be changing and it might now be possible to specify special static dissipative coatings for the vessel and stirrer.

High quality lined vessels, which typically have glass coatings in the range 40–90 mil including a 20 mil groundcoat, may be successively coated with 5–7 layers of glass with intermediate exposures to 20 kV sparks followed by refirings to reduce the number of trapped microscopic bubbles, a likely precursor to pinholes. It is common practice to repair pinholes with special tantalum plugs of various sizes which may range from about 3/16 to 4 in. diameter. These plugs may not be effective in dissipating static, especially where liquid conductivity is very low and the charge is strongly held in an electrical double layer with the wall. Thus, the number of plugs can multiply with continued operation until it is necessary to recoat. Also, plugs are of limited use on impeller tips and leading/trailing edges (where pinholes are frequently observed) particularly where a solid phase is involved (5-4.5.1). Sometimes a sacrificial PTFE boot is used in such cases although if metal contamination is intolerable a spare impeller should be available for anticipated repair downtime. Typically after a batch campaign, the presence of pinholes can be manually determined using commercially available "holiday detectors" comprising either high voltage spark testers (adjustable from 0–30 kV depending on breakdown voltage of coating) or, for coatings less than about 0.5 mm thick, low voltage "wet sponge" testers (typically 9–90 V). Use of the testers requires a clean and dry vessel suitable for vessel entry. An alternative nonentry technique uses an "enamel tester" comprising a conductivity probe to detect metal contacts. In this case the liquid in the vessel must be conductive so the technique may not be suitable for continuous monitoring. Where applicable and practical, the more conductive liquid should be added prior to stirring or other measures taken to increase liquid conductivity (5-2.6 and 5-4.5.2). Where nonconductive flammable liquids or slurries are splash-mixed or stirred, and in recrystallization operations involving nonconductive solvent, the vapor space should be inerted owing to the possibility of static discharges at the vapor–liquid interface. Slower stirring speeds may help reduce pinhole damage but might not be effective without additional measures.

5-4.5. Liquid–Solid Mixers, Blenders and Reactors

Agitators should be covered with sufficient depth of liquid to minimize splashing, or operated at a reduced speed until sufficient depth has been built up. In cases where hazardous static accumulation in the liquid cannot be avoided using the measures discussed here, ignition hazards can be mitigated by inerting plus avoidance of metal projections from the vessel head such as thermowells that could act as discharge electrodes above the charged liquid surface.

While it is rarely possible to add solids first, in certain cases it might be possible to reduce flammability hazards by planned sequential additions of solids and liquids, the objective being to avoid additions of easily ignitable powders where the solvent vapor exceeds about 50% LFL. Hybrid mixtures are discussed in 6-1.3.

Ball mills are sometimes used to simultaneously grind powders and mix them with flammable liquids. Fires have been reported in porcelain ball mills where the flammable liquid was nonconductive, such as toluene or xylene. Charge accumulation is easily predictable in such cases owing to the high surface area of the balls, rapid separation of surfaces, and very small rate of charge dissipation. Since all internal mill surfaces (porcelain) are nonconductive, brush discharges from the surfaces of the balls are to be expected. Although liquid conductivity might be increased as described in 5-4.5.2, adverse conditions such as accumulation of isolated conductive patches on the ball surfaces could offset this mitigation. Inerting should be considered for these operations.

5-4.5.1. Control of Static During Solids Additions

The most frequent cause of static ignitions in these operations is addition of solids to the container. Even if the container is inerted, a sudden large addition of solids may entrain air into the container while expelling flammable vapor from it; a specific example is the Flexible Intermediate Bulk Container (6-7). Also, the sudden addition of a large volume of solids might result in static discharge from a floating pile of charged powder. It is recommended that direct addition of solids be done only in 50-lb (25-kg) batches from effectively groundable bags or fiber drums. Larger batch additions should be done via an intermediate hopper with rotary valve, or its equivalent. This hopper can be separately inerted to reduce air entrainment into the mixing vessel while expulsion of flammable vapor into the operating area can be avoided using a vessel vent line to a safe location. Solids additions from nonconductive plastic bags can be hazardous even if the solids are noncombustible, such as silica. Many fires have resulted from bringing plastic into the vicinity of a vessel opening. Bags should be constructed of paper, plied paper–plastic in which the nonconductive plastic film is covered by paper on both sides, or antistatic plastic (see also 6-6). Since grounding clips may be impractical, such bags may be effectively grounded by contact with a grounded conductive vessel or skin/conductive glove contact with a grounded operator. Fiber drums should not have a loose nonconductive plastic liner which might leave the fiber pack and behave as a plastic bag. The metal chimes should be grounded. Personnel in the vicinity of an opening on a vessel containing flammable liquid should be grounded and special attention should be paid to housekeeping, since accumulation of nonconductive resin or lacquer on the floor or items such as grounding clips may destroy electrical continuity.

5-4.5.2. Control of Static during Stirring

While solids are being dissolved or emulsified in nonconductive liquids the rate of charge generation can be large depending on factors such as solids loading, particle size and agitation rate. Dissipation of the charge is frequently achieved by raising the conductivity of the continuous phase with addition of a large fraction of conductive liquid, for example, 20 vol% of a ketone might be added to a hydrocarbon (Figure 5-4.5.2). The liquid conductivity should be greater than 5000 pS/m and up to 10,000 pS/m if possible. Suitable conductive liquids are typically referred to as "oxygenated sol-

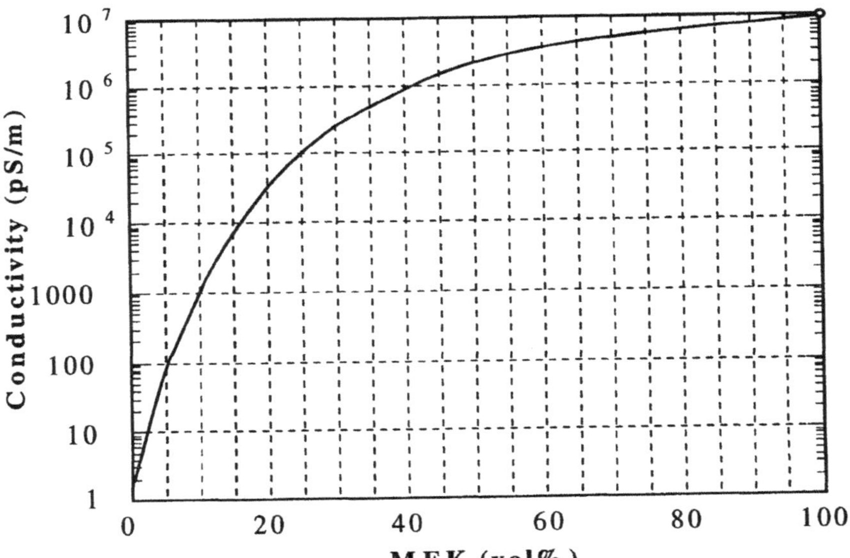

FIGURE 5-4.5.2. Effect of added methyl ethyl ketone on conductivity of a toluene sample.

vents." Note that esters may be "semiconductive" and less effective than alcohols or ketones. Changes in flammability hazard of the resulting mixture must be evaluated. For example, the addition of MEK to xylene may greatly increase the ignition hazard owing to the lowering of flash point. An alternative method of increasing conductivity is to use a few ppm of antistatic additive, which has no effect on flash point. This approach may require conductivity evaluation in the process to ensure that the additive is not removed by suspended or emulsified solids. The process change may also require reapproval of the product by regulatory agencies. Metal-free antistatic additives are discussed in 5-2.6. Although ignition hazards can alternatively be controlled by inerting, charge accumulation may result in "pinhole" lining damage to equipment such as enamel or glass-lined reactors (5-4.4.1).

5-4.6. Vacuum Trucks

Under vacuum conditions the flash point of liquid in the cargo tank is reduced (5-1.1.3) although vapor accumulation in the truck tank is offset by air throughput. Single component volatile liquids having a vapor pressure exceeding about 80 mmHg at 20°C should not be picked up using a vacuum truck because they might boil, creating downwind air pollution and possibly a vapor cloud hazard. Boiling is an obvious hazard for Class 1A liquids but has also been found to be a problem for Class 1B liquids, the more volatile of which (such as acetone) may be substantially lost by evaporation. In the case of mixtures, the listed vapor pressure may be due to components representing only a small fraction of the bulk liquid and a small amount of boiling may be acceptable provided the vapors are discharged to a safe location; also, the more volatile fractions may have weathered off by the time pick-up is scheduled.

Most "static" incidents appear to have been due to improper hose selection and fires/explosions typically do not involve cargo tanks. Two mitigating factors during vacuum truck loading result from the low pressure inside the cargo tank, which can run as low as 50–80 mmHg. The first is that gas MIE increases as pressure decreases (Figure 5-1.4b). The MIE of a typical aliphatic hydrocarbon at pressures less than 0.2 atm (150 mmHg) exceeds the effective energy of brush discharges. Therefore, while loading flammable liquids, there is unlikely to be any static ignition source inside the cargo tank provided all metal components are properly bonded together. Also, small metal items that are inadvertently sucked up should have too small a capacitance to cause ignition. Low MIE vapors such as carbon disulfide may be exceptions since a fire was reported involving a carbon disulfide–water emulsion. The second mitigating factor is the small overpressure that will result even if a deflagration does occur during loading. If a typical maximum 9:1 pressure ratio is developed, the maximum pressure starting at 80 mmHg will be roughly the same as the ambient pressure outside the tank. Owing to these mitigating factors plus practical considerations, flow rates do not need to be restricted to those recommended for road tanker operations (5-4.6.1).

Hoses must be chemically compatible with all materials that may be picked up. For static control, hoses should be conductive or semiconductive so that metal wands and hose connectors are bonded to the grounded truck via the hose. For grounding requirements see 5-3.3.1. In no case should plastic wands be used. If a metal clamp is used to hold a wand in place it should be bonded or otherwise electrically continuous with the grounded hose. A drum explosion due to the use of a plastic dip tube and connected rubber hose is described in [8]. A conductive hose should preferably be a 100% flexible metal type. An alternative is to select a semiconductive hose so that loss of continuity in a bonding spiral does not isolate a wand or create spark sources along the hose (5-3.3.1). However, owing to the possibility of sucking up abrasive solids a thin semiconductive hose lining might be unsuitable. Where practical, the truck should be grounded either by bonding to grounded equipment or via a grounding rod (4-1.5). A fire in a toluene sump was caused by a spark from a spiral-wound conductive hose which penetrated about 2 mm of rubber hose material to the metal rim of the sump; although the spiral was bonded to the truck, the latter was not grounded [33]. Ideally, the electrical continuity or individual grounding of all conductive components (truck tank, conductive hose, metal connectors, wand, containers being emptied) should be established by test. During loading or offloading flammable materials, personnel should not stand in front of or behind rear open-dome type trucks. These are designed so that the rear of the truck tank is the weak point and this may act as a relief vent if deflagration occurs inside the truck. Personnel wearing conductive or antistatic gloves will normally be grounded via the metal wand, but special consideration should be given to personnel grounding where vacuuming is done from opened equipment or other confined area where flammable atmospheres may rapidly accumulate.

5-4.6.1. Additional Precautions and Alternatives

Some liquids might create an ignitable froth layer (5-1.3.1), although mist accumulation will be offset under conditions where there is periodic suction of air through the hose. It is possible to specify a vacuum truck tank with MAWP > 50 psig, sufficient to contain a deflagration beginning at up to atmospheric pressure. This will not necessarily prevent hazards due to flame or hot gas venting from the tank vent or via the hose, but will prevent explosion of the truck tank.

Where "flammable" liquid spills (5-1.1) are to be collected, the use of a suitable pump (diaphragm or gear pump) and tank truck might be considered as an alternative to the vacuum truck. This allows better control of operational variables; where nonconductive flammable liquids are to be picked up, excessive flow velocity, and the presence of a second phase such as undissolved water or entrained air can all lead to high rates of charging in the hose. Pumping to "rear open dome" vacuum trucks should be avoided since owing to the large dome area, even a small positive pressure in the cargo tank can cause the latches to fail. Another alternative to direct loading to the cargo tank of a vacuum truck is a cyclone separator attachment which loads to steel drums. Drums and associated conductive equipment should all be electrically bonded to the grounded truck.

At the time of writing, API Publication 2219 "Safe Operation of Vacuum Trucks in Petroleum Service" was in preparation.

5-4.7. Plastic Tanks

For portable tanks see 5-7. Where possible nonconductive plastic tanks should be avoided for flammable liquids. All plastic tanks present external ignition risks if their outer surfaces become charged. Also there is no Faraday cage shielding of internal flammable vapors from the effects of lightning as in the case of metal tanks. Where lightning and other risks are unacceptable a lined metal tank should be considered. An intermediate remedy is to use a plastic tank that has a grounded, conductive mesh buried in all external tank walls, although this might not be adequate for direct lightning strikes. In 1997, a plastic tank containing HCl accumulated a hydrogen–air atmosphere which ignited during an electrical storm, causing the tank to travel 300 ft. laterally and to reach an elevation of about 250 ft. From inspection of tank heat damage, a direct lightning strike to the tank probably did not occur. However, a lightning strike map (using the "National Lightning Detection Network") showing probable strike zones during the storm indicated several strikes whose 97% confidence ellipses included the tank.

5-4.7.1. General Mitigation Measures

The use of fixed plastic tanks is especially problematic for nonconductive flammable liquids because the efficiency of grounding aids (such as a conductive dip pipe or a submerged grounded metal plate) decreases as liquid conductivity decreases. At very low conductivities of around 1 pS/m or less such measures may become less effective owing to the slow rate of charge migration to the grounding system. Also, since charged liquid inside tanks may induce charge on conductive objects either inside or outside the tank, inerting will not remove external ignition hazards from the area. Where plastic tanks must be used for such liquids, careful attention to grounding around the tank should be carried out and inerting should be used where possible.

Where plastic tanks are used for conductive or semiconductive liquids in the presence of flammable gases, bottom filling using a conductive dip pipe will greatly assist charge relaxation inside the tank. A slow start (5-3.4) should be used until the dip pipe is submerged. A slow start can be avoided using an internal grounding plate on the tank floor (5-4.7.2). Where corrosive liquids such as acids are involved it is important to consider long-term compatibility with the grounding system. An explosion of a waste acid tank occurred after a broken dip pipe caused splash filling, which in turn led to static accumulation in the tank and a spark which ignited hydrogen gas. In this case the potential for hydrogen generation was previously unrecognized. Following the event the cause was immediately obvious and a qualitative hazard analysis should have prevented the incident.

5-4.7.2. Use of Internal Grounding Plates

It was theorized that the charge flowing into a plastic tank resting on the ground, or other external grounded surface, would predominantly reside in the capacitor formed across the tank floor [173]. In the absence of large conductive surfaces external to the tank sidewalls, their capacitances would be small compared with the floor and only a small fraction of the total charge flowing to the tank would reside there. It followed that a sufficiently large grounded plate situated on the tank floor should be effective in dissipating charge even though only a small fraction of the internal wall area would be grounded. Tests were conducted using a 12 m^3 fiber-reinforced polyester (FRP) tank having a 2×2 m floor area. A 1.0×0.5 m steel plate or heavy steel mesh was found to be capable of removing a large fraction of the charge from flowing low-sulfur diesel oil having a conductivity of about 20 pS/m. The oil was charged using a high voltage diode injector device. For charge densities of up to 185 $\mu C/m^3$ and flow rates up to about 300 gpm, the current collected by the plate or mesh at a given liquid depth was directly proportional to the current flowing to the tank. Observed current collection efficiencies were 30–70% during filling. The highest values occurred at small liquid depths (about 0.4 m) prior to submergence of the side-entry filling pipe. After complete submergence of the filling pipe at about 1 m depth, the collection efficiency was about 50%. The efficiency gradually decreased at larger depths due to the increasing capacitive effects of countercharges on the external tank walls [206]. The basic experimental arrangement plus a discussion of finite element computer modeling of the results is given in [174].

For flat-bottomed plastic tanks resting on the ground the plate or mesh can be placed directly beneath, and bonded directly to, the grounded metal dip pipe. However, the distribution of charge in an isolated plastic tank depends on the relative location of all external conducting objects, not only the ground. Where a plastic tank is elevated and significant capacitance due to adjacent conductors appears at side walls, the use of additional grounding plates on the walls concerned might be considered. The effect of wall capacitance is most severe if the tank is equipped with a grounded, conductive grid buried in the side walls. However, this is only of importance regarding internal grounding measures and the principal effect of such grids is beneficial. The grounded grid is a barrier to electric fields, so that charge inside the tank cannot cause hazardous induction outside the tank.

Exxon carried out test work in a horizontal, cylindrical FRP tank roughly 3.1 m long and 1.2 m diameter [207]. Tank volume was 3.6 m^3 (nominally 1000 gallon). Kerosene jet fuel with a conductivity of 6–15 pS/m was charged using a microfilter and flowed into the tank via a 3-in. metal pipe positioned 0.6 m from one end. It was found that 50–90% of the current generated in the filter was removed by the fill pipe assembly, although external discharges occurred frequently to conductive objects around the outside of the tank. Two sizes of grounded plate (25 × 90 cm and 25 × 168 cm) on the tank floor were then tested. It was found that essentially all of the generated current was collected by the fill pipe plus plate. The outside of the tank was then completely covered with metal foil to simulate the effect of a buried, conductive grid in the tank walls. Such a grid prevents external discharges but maximizes the effect of wall capacitance. Most of the incoming current flowed to the walls, as indicated by the flow of countercharge to the external foil. The currents flowing to the filling pipe, foil and grounded plate were compared as fractions of the current generated by the filter. It was found that the external foil reduced the current collected by the fill pipe and plates to 50–65%, with the larger plate showing slightly greater charge removal. The charge attracted to the walls slowly decayed over a period of the order 1 h. However, static discharges from the liquid surface ceased to be produced a few minutes after filling the tank. The tests showed that grounding plates are effective for plain tanks, tanks with grounded metal grids in the walls and (by inference) buried tanks.

Based directly on conclusions given in the Exxon study [207], API 2003 [3] recommends that the area of metal grounding plates should be not less than 30 in.^{2} per 100 gallons or 500 cm^2/m^3. BS 5958 [2] recommends that for tank volumes up to 5 m^3, the plate area should be 400 cm^2/m^3 and that the liquid should at no point be more than 2 m from the grounded plate. The language in BS 5958 implies that the tank volume and liquid distance restrictions are simply cautionary statements based on the size limits of experimental tests.

5-5. Sampling, Gauging, and Analysis

For all manual operations see 4-3.1 "Personnel Grounding."

5-5.1. *Sample Container Cord*

Grounded stainless steel cord or its equivalent is recommended for sampling and gauging operations. Chains are not electrically continuous and should not be used in flammable atmospheres owing to possible creation of spark gaps. Cord made from synthetic material or cellulosic fiber has the disadvantages that (1) it does not allow proper bonding of conductive sample containers and (2) there is a possibility of charging if the cord slips rapidly through gloved hands. Although natural cellulosic fiber cord will not charge significantly via the second mechanism and should in principle give a small enough RC time constant [Eq. (2-3.8)] to allow charge to dissipate from a sample container, commercial cellulosic cord is frequently composed of a natural–synthetic mix. In practice the risk of static discharges from charged, nonconductive cord is low unless ignition sensitive gases such as hydrogen are present. Tests using charged plastic tubes showed that the ignition probability of atmospheres having a MIE above 0.17 mJ only becomes negligible for tube diameters 5 mm or less [26], suggesting that a greater and usually unrecognized risk is due to nonconductive gas sampling tubes that are rapidly lowered into tanks through gloved hands. Antistatic tubing is available for these applications (5-5.4).

5-5.2. *Sampling*

Sample containers are discussed in 5-8.4.2. For tanks containing a flammable atmosphere this is most safely accomplished using a sampling well, also known as a gauge well. This is a metal pipe, sometimes perforated, through which a sample container or gauging device is lowered into the liquid. The metal pipe is permanently fixed and electrically bonded to the top and bottom of the tank. Since the pipe acts as a Faraday cage, the sample container cannot intensify the electric field inside the tank and create static discharges. Sampling can therefore be done at any time. Where sampling wells are not used, sample containers should not be lowered into tanks while any energy is being added to the liquid, such as by pumping. Unless a sampling well is used, a wait time should be allowed for charge to dissipate (5-4). For stationary nonconductive liquids in large fixed tanks and ship compartments the wait time should be at least 30 min for nonconductive liquids and at least 10 min for semiconductive liquids. Shorter wait times are adequate for road tankers and rail cars (5-4.2). Since charged mist may persist in the tank long after charge has relaxed from the liquid, even conductive liquids may require a wait time if splash-filling or other mist-generating factor is present. The sample thief, or other device such as a sample bottle cage, should be bonded to the grounded tank or other established grounding point. Where atmospheres contain ignition sensitive gases such as hydrogen, metal combinations capable of producing Galvanic sparks, such as copper wire on a galvanized steel tank, should be avoided.

Manual sampling should be carried out by grounded personnel, preferably using grounded metal equipment such as a metal sample thief bonded to the tank via metal cord. A satisfactory alternative is to place one or more glass sample bottles in a grounded metal cage (5-8.4.2).

5-5.3. Gauging

An option is to use noncontact ultrasonic gauges, or electromagnetic radiation devices such as radar gauges, avoiding moving parts and possible loss of electrical continuity. Electronic gauges with data-averaging features can be especially useful where the liquid surface sloshes around, as in ships' tanks or stirred tanks. Gauge wells (5-5.2) are recommended for manual gauging operations; if these are unavailable an alternative to retrofitting is to use a wait time equal to that used for sampling (5-5.2). Manual gauging should be carried out by grounded personnel using grounded metal equipment.

Flotation devices for automatic gauge tapes should be antistatic or conductive with electrical continuity via the tape device to a proper ground. Isolated conductive components must be avoided. Consideration was given to possible ignition hazards of automatic tank gauges comprising a conductive float held between two grounded metal guide wires and suspended by a perforated metal gauge tape [47]. It was concluded that properly constructed and maintained devices are unlikely to lose ground continuity and, further, the grounded metal guide wires greatly depress the electric field acting on the float.

5-5.4. Portable Flammable Gas Analyzers

Even if the device has been properly calibrated for the particular gas in air, some analyzers such as catalytic bead types will not function in oxygen deficient atmospheres containing less than about 10 vol% oxygen. Some portable analyzers are supplied with antistatic tubing but others are supplied with nonconductive tubing such as Tygon.

If a metal wand plus a short length of tubing is used for manual sampling inside tanks or other enclosures, the tubing should be antistatic or conductive to avoid isolating the metal wand, and the analyzer should be either grounded or hand-held by a grounded operator. The hazard of long, nonconductive sample lines charged by slipping through gloved hands is analogous to that of synthetic cord, except that owing to the larger typical diameter of sample tubing the ignition risk is greater (5-5.1). When sampling large tanks for gas-free status, it may be required to sample near the base of the tank. In this case it is especially important that if a metal object is used to create a plumb bob it is not isolated from ground by a nonconductive sample line. In both these cases, antistatic or conductive sample tubing should be used; tubing approved by the American Bureau of Shipping and the US Coast Guard is commercially available. ABS/CG requirements are based on a "Safety of Life at Sea" (SOLAS) regulation calling for "Anti-Static, Electrically Conductive" tubing. However, for portable analyzer applications the tubing does not need to be conductive. Antistatic tubing is sufficient to dissipate charge accumulated on the tubing itself. When using antistatic tubing with a molded-in wire, the wire should be bonded to the analyzer and to any metal probe (wand or plumb bob) inserted into a potentially flammable atmosphere.

5-6. Tank Cleaning

5-6.1. Water Washing

During rapid shear of water–air interfaces such as during spray washing, thin films and filaments are produced from the interfacial layer which coalesce to form fine mist. The interfacial layer contains a net charge owing to the presence of ions, which in pure water are negatively charged (OH^-) ions held by the preferential orientation of water dipoles at the surface. Farther beneath the interface, positive ions do not form a surface bound layer since they are thermally randomized, so coarser droplets formed from the body of the water jet carry a net positive charge. The coarser droplets settle out faster leaving behind fine mist containing net negative charge [85]. As washing progresses, a steady-state is eventually reached in which the mist tends to achieve a constant charge density, the rate of production being countered by mist removal and charge neutralization by corona. If salt water, detergents or other contaminants are present, the fine mist may instead be charged positively owing to a reversal of the initial interfacial polarization. Hot and/or cold washing cycles using portable or fixed washing equipment therefore results in a cloud of charged water droplets which creates an electric field inside the tank. Hot washing creates higher charge densities and potentials than the cold wash.

5-6.1.1. Conductive Tanks

Provided all conductive components in the tank are grounded the only source of a static spark is charged, isolated slugs of water. These may form during breakup of a jet, drainage from an overhanging tank component, or sloshing of water heels inside a vessel at sea. In the absence of steam, test work and calculations (A-5-6.1) suggest significant hazards only occur in tanks exceeding 1000 m^3. Precautions for washing large tanks such as marine center tanks are given in ISGOTT [5]. Analyses (A-5-6.1) suggest that the hazard can be neglected for tanks less than 100 m^3, although more conservative guidelines have been published for industrial tank applications [127] as noted below.

There is negligible "isolated water slug" hazard for barge tanks owing to their size and geometry, provided steam is not injected directly to the tank during any hot wash, for example during start-up of a steam assisted hot wash system [186]. Consideration might be given to a cold wash prior to hot water washing, since this may in some cases reduce the flammable vapor concentration with associated benefits in controlling air emissions.

Conservative guidelines for industrial tanks containing flammable atmospheres have been published by a group of four major European companies [127]. Washing is restricted to continuously drained metal or enameled metal tanks having a volume no greater than 30 m^3 (or diameter no greater than 3 m) with all conductive parts grounded. The sprayed water has a maximum feed rate of 300 L/min and pressure does not exceed 500 bar. These guidelines are based on the assumption that incendive brush discharges occur in tanks where the breakdown field of air is exceeded at the grounded spray head, the latter being modeled as a 4-cm-diameter grounded electrode placed in the center of a charged cloud of droplets with a volumetric charge density of 240 nC/m^3. The requirement for a water slug is omitted. The tendency of wet surfaces to produce corona rather than brush discharges is not considered (A-5-6.1).

On June 3, 1998, in Rotterdam a 165-m^3 cone roof tank containing methyl tertiary butyl ether (MTBE) exploded while being washed, causing a fatality and launching the 1.4 tonne roof a distance of 100 m. The tank was 10 m high by 4.5 m diameter. Initially the tank contained 27% MTBE vapor with the balance nitrogen. However, as washing proceeded a vacuum truck was used to suck sludge, water, and residual MTBE out of the tank bottom. As the vacuum hose lost contact with the liquid layer, the resulting negative pressure intermittently sucked air into the tank via the top manway, which held a high pressure washing device. Turbulent mixing by the washing jets helped form a flammable mixture in the tank. The explosion occurred during the fourth cleaning cycle and the fatally injured man was blown from the roof shortly after adjusting the washing device. Two "static" mechanisms considered were a static discharge involving the water mist and a static discharge from an ungrounded vacuum hose. The latter would have to be ruled out before attributing this incident to a charged mist mechanism involving brush discharges or charged water slugs. The company involved had previously ventilated their tanks to the air to reduce flammable vapor concentrations to a safe level, but this approach had been abandoned owing to local emission constraints.

5-6.1.2. Nonconductive Plastic and Plastic Lined Tanks

Where flammable gas/vapor is present or may be generated during cleaning it is recommended that these tanks be inerted. For example, a deflagration occurred during water washing of a lined internal floating roof tank and all causes other than a spark from charged water in the tank were ruled out by the oil company concerned. Where sensitive gases such as hydrogen may be generated the risk of static ignition increases (5-6.4).

5-6.2. Solvent Washing

The charge densities of mists created by flammable and combustible solvents are similar to those from water washing and similar precautions should be taken regarding grounding of conductive components. Where possible the tank should be inerted. Since it is not usually possible to control the flammable atmosphere by operating below the LFL, solvent washing should only be used in noninerted tanks if water washing with added cleaning agents such as detergents, or water washing followed by steam cleaning, are not practical alternatives. If possible the solvent used should have a flash point above the maximum temperature expected for the operation (5-1.1.5). Although this will not prevent the formation of flammable mist in the tank

(5-1.3), it will be less hazardous than using a solvent that creates a continuously flammable vapor atmosphere (5-1.4.4). Solvents having flash points greater than 200°F can often be used in these operations and in general the solvent should be selected at least in part on the basis of high flash point.

If the solvent is nonconductive, additional precautions are needed. These are (1) continuous draining to prevent build-up of charged liquid in the tank and (2) measures to avoid accumulation of solids if the solvent is recirculated. Measures to separate solids before recirculation, such as filtration, may increase static generation (5-3.5).

Expert review should be made of any procedure requiring personnel in the tank during solvent cleaning operations. In 1998, a flash fire fatality occurred in England during manual cleaning of a tank using xylene at less than its flash point. The operation involved an ungrounded metal bucket hanging in the tank from which xylene was taken with sponges. The wringing out of the sponges is believed to have charged the bucket which subsequently sparked to the tank internals, igniting a mixture of xylene plus a previously unrecognized but significant concentration of residual MEK vapor in the tank. The incident was especially serious as fire resistant clothing was not worn and a container of xylene was spilled during emergency egress from the tank.

Where inerting is not used, one guideline restricts operations to metal or enameled metal tanks having a volume no greater than 5 m^3, a diameter no greater than 3 m and all conductive parts grounded. The liquid should contain a maximum of 1 wt% insoluble solids, have a maximum feed rate of 60 L/min and the delivery pressure should not exceed 50 bar. The vessel should be continuously drained to prevent liquid accumulation [127].

5-6.3. Steam Cleaning

Steaming creates high charging currents in the hose, primarily as a result of expansion at the nozzle. All components of the steaming system must be conductive and grounded to prevent sparks. The expansion of wet steam also produces a charged jet whose charge density depends on orifice geometry plus steam temperature and pressure. Ignition tests have been performed to address brush discharge hazards from steam jets. However, these have typically failed to demonstrate ignition even of stoichiometric hydrogen–air mixtures [46, 85]. The mist created by steaming has a large charge density resulting in a correspondingly large space potential which increases with the diameter of the tank (A-5-6.1). Large fields and potentials might result in corona and brush discharges from tank internals or spark discharges from isolated slugs of water. Although steam is usually a more effective inerting gas than nitrogen, this is only true to the extent that the oxygen concentration is reduced to less than the LOC. Credit cannot be taken for condensed mist. Until the tank temperature exceeds 80°C, corresponding to a steam partial pressure ≥0.47 bar, the residual concentration of oxygen derived from air can exceed 10%. Therefore, a flammable mixture may persist long into the steaming process.

As discussed in 5-6.1.1, a conservative guideline for water washing tanks containing flammable atmospheres limits the maximum tank size to 30 m^3. The basis for this provision is to avoid brush discharges at the washing head. If a similar analysis is carried out for steam cleaning, the maximum allowable tank size will be smaller owing to the greater charge densities produced in steam mist. Recommended practices such as [1,2] limit the tank size to 100 m^3 instead. No rationale is given for this round number. In principle, steam cleaning should be most hazardous if charged slugs of condensate are ejected into a large, roughly cubic or squat cylindrical tank containing an easily ignitable atmosphere. If there are no conductive water slugs and all metal components are grounded, the only source of ignition during steaming is the brush discharge. There is no direct evidence that brush discharges are a hazard during steaming although the probability of their occurrence increases with tank volume. The user should consider risk tolerance where tanks up to 100 m^3 are steamed while containing a flammable atmosphere.

Where steam cleaning of large tanks is carried out, gas freeing or inerting should be considered. Gases initially present in the tank can be removed by ventilation or water displacement. During steaming, flammable vapors may be released from liquid heels or deposits. An option is to periodically test for flammable vapor and either ventilate or (if the vapors are water soluble) perform water washing until the vapor concentration is less than 25% LFL. Alternatively, steaming may be carried out using inert gas to stay below the LOC. Unless a tank is designed for vacuum, vent capacity must be sufficient to prevent underpressure damage as steam condenses.

5-6.4. Acid Washing

These operations often generate hydrogen and other flammable gases. Hydrogen is commonly produced and is a special hazard owing to its very low MIE, wide flammable range and fast burning velocity. Further, hydrogen flames can be practically invisible in daylight, creating an additional injury hazard. Static has caused numerous explosions. In one case, a reactor containing a cobalt catalyst residue was cleaned and the acid wash was transferred to a polypropylene portable tank via a conductive hose. Hydrogen was generated via acid–catalyst reaction. The portable tank was splash-filled, which allowed the waste liquid to accumulate static charge. When the charged liquid surface reached the conductive hose hanging inside the portable tank a spark was generated, igniting the hydrogen–air mixture. An operator was injured by the subsequent explosion. Hydrogen continued to be generated in the reactor and burned for some time afterward as it exited the hose. Where possible these operations should be inerted (NFPA 69) and careful attention should be paid to grounding. Plastic tanks and tanks with nonconductive linings require particular care since in the absence of a suitable grounding system even very conductive liquids such as water or acid rinses can accumulate charge. Conductive liquids typically produce sparks instead of brush discharges; hydrogen may be ignited by spark at a potential as low as 100–300 V (Table A-1-4.1.3b). To put this in perspective, ignition of gasoline vapor in air via brush discharge from a nonconductive liquid requires a minimum surface potential of about 25,000 V (A-5-4).

5-6.5. Grit Blasting

Where possible tanks should be gas free or inerted. Conductive or semiconductive hose should be used and attention given to personnel grounding where flammable atmospheres might exist. All conductive components, especially the hose nozzle, should be grounded to avoid static sparks. In addition to static hazards, the possibility of uncovering pyrophoric iron sulfide deposits should be evaluated. It is highly improbable that ignitable dust clouds can be produced as a direct result of using organic grit such as walnut shells.

5-7. Portable Tanks

As defined in NFPA 30 portable tanks have capacities above 60 gallons (227 L) and usually below 660 gallons (2498 L).

5-7.1. Metal Portable Tanks

Bottom filling should be used for nonconductive flammable liquids and filters should be placed at least three relaxation times upstream (5-3.4 and 5-3.5). The container should be grounded prior to opening and sealed prior to removing the grounding system. Filling rates should be similar to those used for drumming (about 60 gpm or less) unless the container is inerted. If the dip pipe does not extend close to the bottom and the vessel is not inerted, a slow start of 1 m/s or less should be used until the dip pipe is submerged to about 6 in. Conductive and semiconductive liquids present minor static hazards except during operations such as sampling and gauging where equipment or operators in the vicinity of a tank opening are not grounded. Tanks with nonconductive linings present hazards somewhat more severe than the case of drums (5-8.2.2) owing to their larger volumes and hence larger potentials that may be generated for equal charge densities.

5-7.2. *Plastic Portable Tanks*

See NFPA 30 for general restrictions. This category of tanks includes Intermediate Bulk Containers (IBCs) comprising plastic jugs within metal support frames, and composite IBCs comprising plastic jugs independent from their external metal shells. Although IBCs might be bottom-unloaded, avoiding hazards associated with manual insertion of unloading lines, manual removal and replacement of vent caps should be carried out by grounded personnel. Attention should be given to ensuring safe handling of "empty" plastic jugs containing flammable vapor–air mixtures following unloading, including cleaning and/or disposal.

Tanks manufactured from nonconductive plastic should not be used for flammable liquids unless the tank is inerted prior to filling, and the operation is subject to expert hazard analysis. Such an analysis should include the evaluation of emergency response both for a leak and an external fire. Highly charged, nonconductive liquid might give rise to propagating brush discharges from the plastic in addition to brush discharges from either the plastic or liquid surface. Additional measures for liquids capable of generating flammable vapor atmospheres might include inerting and personnel grounding during unloading. A case history of an explosion in a plastic portable container used for collecting an acid wash is given in 5-6.4.

5-8. Portable Containers Less Than 60 Gallons Capacity

Drums comprise metal, lined metal or plastic containers with capacity up to 60 gallons (227 L) and are the largest containers considered in the following sections. Metal and plastic-lined metal drums can have either fixed or removable heads. It should always be ensured that drums are sealed prior to handling and that when opened they are properly grounded. Since plastic drums cannot be grounded they should not be used for flammable liquids or handled in flammable atmospheres without expert review of the hazards. Although drum explosions due to static have low probability, the large number of operations means that incidents are not rare. Furthermore, the consequences are often serious. Although the volume of a drum is relatively small, a flammable mixture that ignites via static is likely to be close to its most easily ignitable composition. The "most easily ignitable" mixture also creates the highest pressure when it deflagrates inside a drum. Static ignition can result in fire and/or explosion, with potential for impact injury. If a steel drum fails at the bottom chime due to an internal explosion it can rocket for tens of feet, spilling its contents and creating a serious risk via mechanical impact and fire [8]. Missile hazards can result from detached drum heads even if the drum remains in place. As an example, an "open-head" drum containing MEK being rolled on its lower chime exploded and the clamped-on head detached, cutting off the operator's nose. Although full details were unavailable, it appeared that a spark from the operator ignited MEK vapor in the vicinity of an opening in the drum head. For manual loading/unloading operations involving flammable vapor, it is recommended that personnel are grounded so they are not a source of ignition (4-3.1), and the bungs and/or drum lid are tightly closed prior to moving the drum.

5-8.1. All-Steel Drums

a. Filling. See 4–1 for general grounding requirements. Drums should be grounded and bonded to associated conductive equipment such as weigh-scales and filling lances. Simple frictional contact between a filling nozzle and the drum port might not give a satisfactory bond especially if a plastic bushing or paint layer is present at the drum port. Proper bond connections should be made using clamps capable of penetrating any paint layers on the drum, typically screw types or spring clamps employing hardened metal points. Bond connections on the upper chime should be located away from the drum ports prior to bung removal and the connections should not be removed until the bungs have been replaced. Braided and uninsulated stainless steel bonding cable is preferred. If insulated cable is used without a ground indicator, a broken cable can go unobserved between scheduled tests. Any filter in the filling line should be grounded and where possible placed at least three relaxation times upstream of the filling pipe outlet unless the drum is inerted prior to filling. If neither option is practical for nonconductive liquids, the filter should be placed as far upstream as possible and reduced flow rates should be used [8]. For conductive liquids either splash filling or bottom filling is acceptable. The filling rate for noninerted drums should be such that the drum takes 1–2 min to fill (for example, ≤30 gpm with a 1-in. filling pipe and ≤60 gpm with a 2-in. filling pipe); as a reference, gasoline pumps typically dispense at 10–12 gpm. Filling/emptying lances should be conductive and grounded, and when used for nonconductive liquids should preferably have pointed tips such as a 45 degree cut to minimize the probability of a brush discharge [8]. Unless inerting is used, automated filling devices should not be programmed to suspend blunt lances above nonconductive liquids as a solution to dripping from the wetted lance between fills, since this practice may result in brush discharges from the liquid [8].

b. Dispensing. Pressurized gas should not be used to dispense liquid. Drums should be grounded and the use of a ground indicator, possibly interlocked, should be considered for drums not dispensed from racks. For gravity dispensing from racks, drums should be fitted with spring loaded metal valves which shut automatically when hand pressure is released. Each drum should be equipped with a drum vent, typically at the larger bung; 2-in. dual-action safety drum vents incorporating a flame arrester are commercially available. Dispensing should normally be via the smaller (¾ in.) bung unless viscosity considerations do not allow this. Small conductive containers such as cans should be bonded to the rack. For vertical dispensing, drum pumps and dip pipes should be conductive and grounded. The pump must be suitable for the service. For flammable liquids, plastic hand pumps should not be used and instead these should be of groundable metal construction. Hand pump designs incorporating an internal flame arrester are preferred. Hoses should either be semiconductive or conductive and grounded (5-3.3). To reduce vapor emission the dip pipe should fit closely into the 2-in. dispensing opening. As liquid is dispensed, air will be sucked into the drum primarily through the vent opening. Use of a flame arrester might be considered for drums used for occasional dispensing over a long period. If vacuum unloading is done, additional measures such as anti-siphon devices and pressure interlock with any inert gas supply might be needed.

c. Cleaning. Drums should be grounded prior to opening for cleaning operations such as steaming. Cleaning equipment should be grounded. Until certified gas-free, drums should be assumed to contain flammable vapor and be grounded unless closed. An explosion occurred while discharging steam via ¼-in. nipples into an "empty" inverted styrene drum on a wooden platform, tossing the drum 10 ft. in the air and causing a fire [31]. Even a new, unused drum might contain vapor following solvent wiping during manufacture and a drum explosion has been reported due to a label being welded onto such a drum [8].

5-8.2. Plastic Lined Drums

Linings inside metal drums hinder the flow of charge to the grounded metal shell and present a more hazardous case than all-steel drums; in principle, even conductive liquids can accumulate charge. However, the lining has a relatively large capacitance which provides considerable mitigation compared with an all-plastic drum. If the lining is assumed to be a perfect insulator over the timescale of loading, and the liquid is splash-filled so no charge is lost via a dip pipe, the charge must flow to the insulated wall and reside there forming one plate of a large capacitor. The external grounded drum wall provides the other capacitor plate. Given the reasonable assumption that most of the charge migrates from the bulk liquid to its surfaces, the surface potential of the liquid in the drum V is approximately equal to Q_T/C, where Q_T is the total charge and C the capacitance of the floor and the walls. The equilibrium potential V at any time is directly proportional to the charge density entering the drum Q_V and the thickness a of the lining [8]

$$V = Q_V Ftab/[\varepsilon_r \varepsilon_0(\pi b^3 + 2Ft)] \tag{5-8.2}$$

where F = volume flow rate (m^3/s), t = elapsed filling time (s), and b = drum radius (m)

Static effects of thin internal coatings such as phenolic paints can be neglected even if they are applied so that there are no pinholes. Their resistivities are orders of magnitude less than polyolefin lining materials, plus the coatings are typically so thin (0.5–1 mil) that the associated high capacitance and small breakdown voltage prevents large surface potentials from being attained (5-4.1.3). Liquid surface potentials of 10–15 kV were developed when nonconductive kerosene was charged to about 570 $\mu C/m^3$ by a bag filter and directly splash-filled into "poly steel" drums containing molded polyethylene liners 1.8–2.8 mm thick [8]. This is less than the threshold potential for incendive brush discharges. However, since sparks from more conductive liquids can be incendive above about 1 kV it is recommended that bottom filling using a grounded metal dip-pipe be used for all drums containing molded nonconductive liners. Drum grounding requirements and filling rates are otherwise the same as for all-steel drums.

The probability of a propagating brush discharge (PBD) in a plastic lined drum is negligible. A typical 55 gallon drum has a diameter of 0.572 m and filled height of 0.811 m, so the wetted area is 1.71 m^2. In order for a PBD to occur, a minimum surface charge density of 2.5×10^{-4} C/m^2 is required (2-6.5) corresponding to a total charge of 4.29×10^{-4} C over the internal drum area. If the drum is splash-filled and the liner is a perfect insulator, so that no charge can dissipate, the liquid entering the drum would require a volume charge density of 2060 μC/m^3. Such a large charge density is not credible unless the drum is splash-filled directly from a filter, disregarding recommendations concerning dip-pipes and filter placement. Additional mitigating factors are that PBDs are only an ignition hazard above the liquid surface and the assumed PBD charge density threshold of 2.5×10^{-4} C/m^2 is probably conservative (2-6.2.3).

5-8.3. Plastic Drums

See NFPA 30 for storage restrictions. Since plastic drums cannot be grounded they pose inherent static hazards in flammable atmospheres. The use of plastic drums for Class I flammable liquids is discouraged and this should be considered only in special cases where plastic is essential for quality purposes, such as an intolerance for metal contamination. Where such a case is identified, expert review of the hazards is recommended. Although drums could be inerted prior to loading or filled at a temperature less than the liquid flash point, similar precautions are more difficult to apply to the unloading operation. If the operation is not under the control of the supplier, for example at a customer site, it is difficult to enforce recommendations from the expert hazard review. For Class II combustible liquids, static should be addressed where (1) the liquid may exceed its flash point during filling or emptying or, (2) the drum could be stored or handled in a flammable vapor atmosphere. In case 1, options include bottom filling and cool-down provisions prior to unloading, especially if the drum has been in direct sunlight or in a hot storage area. In case 2, plastic drums should where possible be stored away from drums containing flammable liquids so that the ignition hazard of discharges from the external drum surfaces is avoided (Figure 5-8.4b).

5-8.4. Hand-Held Containers

The fire risk from static ignition increases with the volume and volatility of flammable liquid handled. Hence, the smallest volume container capable of

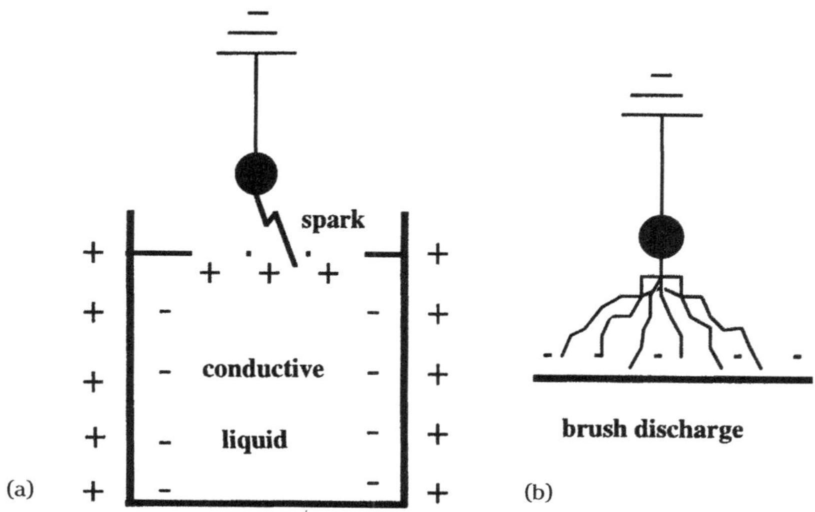

FIGURE 5-8.4. Static discharges associated with plastic containers. (a) Charged container induces charge on conductive liquid which sparks to ground. (b) Charged container exterior discharges to ground via brush discharge

fulfilling a particular need should normally be selected and this should not exceed 5 gallons. Groundable, listed metal safety cans with incorporated flame arresters are preferred, especially those types equipped with a flexible metal dispensing hose enabling the can to be used without a funnel. In the case of nonconductive containers, which cannot be grounded, the maximum quantity should normally be about 5 L (1.32 gallons) for Class IB and IC liquids, and 2 L for Class IA liquids. An exception is gasoline, where approved 5-gallon plastic safety cans have been widely used for many years with no reported increase in static ignition relative to metal cans (5-9.8.2). This is in part due to the rapid establishment of rich (>UFL) gasoline vapor inside the can; these plastic containers should not be used for other flammable liquids without expert review of the hazards. Unlike gasoline, conductive liquids such as alcohols may become inductively charged [9,16] by a charged plastic container and give rise to sparks (Figure 5-8.4a), and the container may in some cases contain an ignitable atmosphere whenever it contains any flammable liquid.

5-8.4.1. Filling and Emptying Small Nonconductive Containers

Subject to capacity limitations described in 5-8.4 it is common to handle flammable liquids in small glass or plastic containers. The following should be considered for frequent indoor liquid transfers of about 0.5 L (approximately 1 pint) or more, such as a small-scale solvent blending operation. Where possible the operation should be conducted in a well ventilated area such as a hood, using a grounded metal catch tray capable of containing the entire inventory of liquid. When filling a container, a grounded metal funnel whose spout extends to the bottom of the container is preferred. This ensures that if charge is induced on the liquid by a previously charged container (Figure 5-8.4a), it will flow to ground via the immersed funnel instead of sparking to the funnel as the liquid level rises. Plastic or glass funnels should be used only where essential for compatibility reasons. Personnel should be grounded.

5-8.4.2. *Containers for Sampling*

Ignition risk is greatly increased when a flammable atmosphere is present outside the small container, for example when sampling directly from a tank or transferring a sample near a manway, since this might precipitate a large fire or explosion. A grounded metal sample thief, or glass bottle in a grounded metal sample cage (5-5) might be selected in such cases. nonconductive plastic bottles should be avoided except where used in well ventilated areas, since they are more easily charged than glass and numerous incidents can be traced to their use. In general they should not be carried in coat pockets or otherwise wrapped in fabric since upon withdrawal a large surface charge can be placed on the bottle. This can cause induction charging and possible ignition when sampling a conductive liquid. To address breakage concerns, alternatives to glass bottles include antistatic plastic bottles and plastic coated glass bottles. In the latter case the plastic coating should be antistatic, which may well be the case owing to the large quantity of plasticizer normally required for PVC coatings. If outdoor sampling is carried out at sample spigots away from tank openings and in freely ventilated areas, and sampled quantities are 1 L (1 qt.) or less, the fire risk is insufficient to require any special procedures other than ensuring that a grounded metal dip tube, extending close to the bottom of the container, is used for filling.

In a demonstration experiment the author placed about 1 in. of water in a 1 pint glass bottle. After drying the outside of the bottle using warm air it was rubbed with a dry paper towel. An induced voltage of 3 kV was measured on the water using a field-mill voltmeter, whose shared capacitance caused roughly a factor-of-two underestimation of the maximum voltage (3-4.1). Glass bottles are relatively difficult to charge by rubbing unless the humidity is very low. It is far easier for plastic bottles to become charged and induce hazardous voltages on conductive liquids inside them.

In 1985, ethylene oxide (EO) ignited in a 1 pint "clip top" glass sample bottle during sampling from a tank car. The ¼-in. stainless steel liquid sample line from the car entered a sampling hood and EO flowed at relatively high velocity (4 m/s) into the bottle through a Teflon tube. The Teflon tube was used to prevent water condensation on the cold sample line, since if water dripped into the sample bottle it would affect turbidity evaluation. EO has a very small ignition energy in air and it was not possible to absolutely rule out ignition by a charged operator. Plausible mechanisms involved (1) EO charging during flow through the Teflon tube, or (2) induced charge on EO liquid from an externally charged sample bottle, or (3) induced charging of the wire clip on the bottle stopper by charged Teflon tube. All of these mechanisms could be avoided by bottom filling with a grounded stainless steel dip tube. The water condensation problem was avoided by fitting the dip tube with a thermally insulating Teflon sleeve, leaving a few millimeters of exposed metal at the outlet. Subsequent to this fire, operators were grounded using wrist bracelets.

5-8.5. Wet-Dry Vacuum Cleaners

If used for flammable liquid spills, which might involve a second phase such as spill control granules or debris, these pose numerous design problems including static, electrical classification, chemical compatibility and industrial hygiene relative to the exhaust. Commercial designs for Class I Group D gas and Class II Groups E-G dust atmospheres are typically air operated via a venturi, so contain no electrical drives. Hoses both for air supply from a grounded air supply outlet, and for liquid recovery, are conductive with additionally static dissipative hose fabric (about 10^5 Ω-m). Filters are also static dissipative. The design is such that all parts are continuously bonded and grounded. Normally the ground continuity is established at prescribed check points before each use. Float or similar mechanisms are employed to cut off suction once the recovery tank has reached capacity level, although additional precautions may be needed to avoid overflow via either siphoning (if the recovery hose is completely submerged in liquid) or if defoaming agents are not used where appropriate. For flammable liquid spills in particular, measures should be taken to ensure personnel are not a source of ignition (4-3.1).

5-9. Miscellaneous Flammable Atmospheres

5-9.1. Clean Rooms

For health care facilities see NFPA 99. Clean rooms require superior static control measures where flammable liquids are handled. Smoke may be intolerable for quality reasons even if a fire is minor. The relative humidity in clean rooms is often controlled at a low value which can decrease the resistivity of surfaces and hence increase static accumulation. Grounded metal safety cans and steel cabinet storage are recommended, taking care to minimize the quantity of flammable liquid in the clean area. Measures include antistatic floors and personnel grounding, including the use of antistatic lab coats. It should be ensured that floor grounding is not compromised by any nonconductive film either from operations or cleaning agents such as floor wax. Careful attention to equipment grounding is required. Plastic equipment should be avoided wherever possible and even small metal transfer containers should be grounded. Items such as cartridge filters require special attention during replacement and if possible the casing should be purged until the element is dry. Disposal of solvent-wet material such as filter elements and towels should be via conductive containers. Such containers should be grounded when in the clean area and sealed before removal. If plastic disposal bags are used they should be conductive or antistatic. Tear sheets generate significant static when they are taken up (5-9.5).

5-9.2. *Water and Steam Curtains*

These do not pose a recognized static hazard if the curtain device and all equipment that could be electrically charged during operation is bonded and grounded [45,46].

5-9.3. *Static Electrification in Gas Flow*

The presence of solids such as pipe scale, or suspended liquids such as water or condensate, creates charged particles carried by the gas phase. Condensate may be created by sudden cooling or expansion of gas through an orifice. An example is sudden release of a pressurized hydrocarbon. Impact of the charged stream on ungrounded objects can create spark hazards. A special case is carbon dioxide, which when discharged under pressure will form charged solid "snow." This can create an ignition hazard if sudden discharge of carbon dioxide is used to inert flammable atmospheres. Static ignition is not a problem with portable carbon dioxide extinguishers used to put out fires. However in some fixed nozzle systems a new flame front could in principle be initiated remote from a flame sensor location, compromising the "barrier" effect of a flame suppression system. Distribution nozzle designs are discussed in [52].

In some cases, gas flow rates are restricted to avoid mechanical ignition via adiabatic compression and frictional impact of small particles with surfaces; examples include oxygen and acetylene. In oxygen systems, ignition associated with particle impact is especially likely just downstream of a location where there is at least a 2:1 absolute pressure drop [50], and may occur at velocities above about 50 m/s. Usually there is negligible static generation in single phase gas flow, although flowing hydrogen has been postulated to react at pipe walls producing an accumulation of dihydrogen cations in the boundary layer, which react rapidly with neutral molecules to give active radical-initiated combustion [43]. This rather speculative mechanism is yet another possible factor in otherwise unexplained ignitions of hydrogen during venting (5-9.4). It is sometimes stated that hydrogen vent ignitions are caused by the inverse Joule–Thomson effect as the gas expands. This is incorrect and calculations show that at 38°C the temperature increase is only about 5°C for an atmospheric release at an initial pressure of 210 bar [212]. Hydrogen and syngas ignition in air might alternatively be explained by entrainment of finely divided, reduced metal oxides, which can be pyrophoric in air. These gases typically ignite when they are vented or leak at pressures above a few hundred pounds per square inch, although the explanation is unclear (5-9.4).

5-9.4. Ignition of Vented Gas

Gases with very low ignition energies (such as acetylene and hydrogen) containing suspended material might be ignited by corona-brush discharges when escaping from stacks at high velocity. This phenomenon is associated with electrical breakdown at the periphery of the charged dust stream being vented, although as noted in 5-9.3, mechanical ignition via friction or compression might occur if air is present in the vent pipe. Static discharges at vent outlets can occur even if the equipment is properly grounded. Atmospheric electricity might also cause "spontaneous" vent fires since the breakdown field of some gases (such as hydrogen) is less than that of air. Hence, the presence of the gas enables atmospheric electric fields to cause gas breakdown and static discharges at lower values than would be required in air. The discharge may be corona-brush type or an upward streamer from the stack that fails to connect with a stepped leader from a thundercloud. Atmospheric electricity can be sufficient to cause static discharges from elevated vents such as stacks even where lightning is not observed locally. Mitigating measures for stacks have consisted of two opposite approaches. In the first (possibly obsolete) method the stack is designed to prevent discharges by using surfaces having large radius of curvature; in [135] a doughnut-like device ("NASA toroidal ring") is discussed for placement on the top of the stack. In the second, the stack is designed to encourage corona discharges by using sharp edges and/or commercially available lightning protection equipment. Where stacks cannot be vented to combustors or flares, they should be equipped with snuffing devices to blow out the flame. Nitrogen or steam at mass fractions above about 0.3 can be used for acetylene, while helium is often preferred for hydrogen vent fires. Snuffing is particularly important for acetylene, which can slowly burn down the stack and, additionally, through any metal flame arrester in the base of the stack. Alternatively, a burn pond can be used to safely dispose of large vented quantities of hydrogen. Other mitigation is discussed in [157].

5-9.4.1. Ball Valve Stem Leaks

The packing in some designs of ball valve can isolate the valve stem or handle from the valve body (5-3.7). Packing leaks may result in ignition via sparks from charged, isolated metal. A particular example is ignition of acetylene. Ball valves are frequently used in acetylene service owing to the sealing action of the packing should a deflagration occur in the line; other types of valve may be sufficiently distorted by the pressure transient to allow flame propagation past the valve even when fully shut. It should be ensured that ball valves have no electrically isolated components, since the vented acetylene has a very small ignition energy and may easily be ignited in air; should acetylene pipework become hot in an ensuing fire, a deflagration or detonation might result from autodecomposition of acetylene inside the pipe.

5-9.4.2. Pressurized Gas Leaks under Insulation

The gas jet may cause impact of loosened insulation material on ungrounded metal weather barriers resulting in charge accumulation and sparking. This mechanism was blamed for ignition of a high pressure ethylene leak from a HPPE product receiver [185]. Isolated metal weather barriers can also create spark sources due to atmospheric electricity. Barriers are most likely to become isolated around line diameter changes such as flanges or valves. Where this is seen as a problem, for example if a decomposable gas is handled, attention should be given to electrical continuity of the barrier. Bonding using separate cables is impractical compared with overlapping and banding of the metal covers.

5-9.5. Hazards of Plastic Sheet and Wrap

Nonconductive plastic sheet and wrap such as pallet wrap present similar hazards to plastic bags. All of these may give rise to brush discharges from their surfaces (Figure 5-8.4b) following rubbing or separation of surfaces. Isolated wet patches may also create spark hazards. An additional problem is charging of personnel during handling. It is recommended that plastic sheet and wrap not be brought into areas that may contain flammable atmospheres. For example, plastic pallet wrap may be removed outside the area and if necessary replaced by a suitable tarpaulin or other temporary cover. Antistatic wrap is available but is usually not an economic alternative. Tear sheets, which are used outside many clean areas, usually generate significant static when they are taken up and precautions are similar to those for plastic sheet.

5-9.6. Oxidant Enriched Atmospheres

The MIE of flammable mixtures containing more oxygen than is available from the air may attain very small values (Appendix B). The MIEs are in some cases so small that normally nonincendive discharges such as corona might cause ignition under process conditions. For most flammable gases and vapors having MIE above 0.2 mJ, corona currents of 200–300 μA have been reported necessary for ignition in air [27,28] and this exceeds most practical rates of charging. However, in oxygen enriched atmospheres, heating of electrode tips by the corona might result in oxidation and additional heating of the metal, possibly forming a hot spot. Such effects might result in ignition at a relatively small corona current, depending on the metal involved and geometrical factors. Ignition under such conditions is difficult to assess since a heat balance for the electrode tip involves losses via convection caused by ionic wind from the corona. Stringent grounding procedures are required to avoid charge accumulation; for general practices involving gases refer to NFPA 53 and 99. Other oxidants such as nitrous oxide and chlorine may similarly reduce the MIE of fuel gases relative to the values measured in air. However, few ignition energy data have been published for such systems.

5-9.6.1. Undesired Oxygen Enrichment

For cryogenic systems see 4-4.1. Where undesired oxygen enriched flammable atmospheres might exist in processes it is common practice to control the flammable atmosphere using procedures given in NFPA 69 in addition to taking measures to control static electricity. Process Hazard Analysis may be required. As an example consider a waste treatment tank using hydrogen peroxide, which normally does not contain a flammable atmosphere. A process upset might be addition of too much peroxide, possibly in the presence of a decomposition catalyst, causing rapid oxygen bubbling. As a reference, complete decomposition of 40 wt% hydrogen peroxide produces 164 times its volume of oxygen at atmospheric conditions. In addition to metal ion sources such as rust, soluble organic decomposition catalysts including enzymes should be considered. In a reactor, an additional factor might be inadequate agitation allowing peroxide to accumulate in the base of the reactor. Another process upset might be introduction of too much organic component in an aqueous feed stream. In this case, bubbling of oxygen through a surface oil layer could result in charged conductive patches which could act as spark sources for oil froth, mist or vapor (5-1.3.1). In such cases numerous ignition sources might exist; for example, organic peroxides might form, concentrate locally and decompose explosively. Mitigation should rely heavily on identification and control of potential upset conditions. Note that vapor space inerting will not prevent combustion of an oxygenated, combustible froth layer created by peroxide decomposition. Also, in some cases sufficient vapor space oxygen enrichment might occur either to defeat vapor space inerting or bring a normally fuel-rich (>UFL) vapor into the flammable region.

5-9.6.2. Ozonization

Silent electrical discharge at up to 15 kV may be used to create concentrations of about 5% ozone in an oxygen stream, which may then be reacted with a flammable or combustible substance for chemical synthesis. Laboratory preparations in (nonconductive) glass reactors have resulted in occasional explosions via static discharges in the oxygen enriched atmosphere, possibly exacerbated by residual vapor space ozone. An alternative to predilution with nitrogen, which forms nitrogen oxides in the ozonizer, is to add nitrogen downstream of the ozonizer. Other measures are to operate well below the flashpoint of any flammable liquid, typically at approximately -70°C using dry ice mixtures, and to select a more conductive solvent, as opposed to a hydrocarbon such as heptane. During shutdown of the system, a suitable inert gas such as argon should be used to thoroughly purge the system. Precautions should be taken to minimize the vapor space volume, avoid tightly closed containers that will not contain the pressure from an internal deflagration/detonation, and take appropriate measures for personnel protection if a flammable mixture might occur during operation. The hazards of unstable peroxides and ozonides, plus materials of construction suitable for oxygen service should be separately evaluated.

5-9.7. Elevated Temperature and Pressure

These conditions generally reduce the MIE of flammable mixtures relative to those measured at or close to ambient temperature and pressure (Appendix

B). They also increase the flammable range, particularly the UFL. An additional effect of elevated pressure is to suppress static discharges (2-6.4.1).

5-9.8. Automotive and Marine

Automobile and marine fuel handling is covered in NFPA 30A. This section addresses specific operations in which static electricity has caused fires and injury. Relevance to the CPI includes employee safety programs and company-owned gasoline fueling depots.

5-9.8.1. Plastic Surfaces

The use of flammable solvents for cleaning can be hazardous. Numerous fires have occurred when using nonconductive solvents such as gasoline to clean surfaces such as fiber reinforced plastic car bodies, resulting in brush discharges. Plastic gasoline cans may also be charged by rubbing, in which case a flammable vapor atmosphere might already be present due to gasoline spillage or an opening on the container.

5-9.8.2. Portable Gasoline Container Filling

These cans have typical capacities up to about 5 gallons. Where practical, metal cans should be grounded during filling and where this is not practical they should be filled when placed on a relatively conductive surface such as concrete rather than asphalt. In no case should they be filled when isolated from ground. Numerous fires have occurred when opened metal cans were placed on fixed or loose plastic bed liners in pick-up trucks prior to filling or moving. Gasoline normally forms a rich, nonflammable mixture inside the can which persists while the can contains a liquid heel, preventing the can from exploding even if it is not fitted with a flash arrester. Hence, fires are generally external to the can. An exception is where a can previously containing gasoline is filled with a high flash point liquid such as kerosene or diesel oil. This is a small-scale example of switch-loading (5-1.4.3). If the can is not fitted with a flash arrester, a flame might flash back and ignite the mixture inside the can, since this may be within the flammable range.

Charging may occur when cans are filled with gasoline or dragged across a rough, nonconductive surface such as carpeting in a car trunk. Sparking may occur between the can and the filling nozzle, a person or the vehicle body. Plastic cans may become charged by rubbing or cleaning but cannot give rise to sparks unless they have metal parts. If the outer surfaces are charged by rubbing, ignition of an external flammable atmosphere (e.g., spilled gasoline) might occur via brush discharge (5-9.8.1). Although plastic containers might not be deliberately rubbed, they can become charged by rubbing against clothing as they are carried. There is no evidence that 100% plastic gasoline cans have a greater incidence of static ignition than metal cans. However, the cans should not be used for conductive flammable liquids such as denatured alcohol (methylated spirits), since the hazards are different. If the can is externally charged, sparks may be possible from inductively charged liquid inside the can. Also, if the liquid has a smaller vapor pressure than gasoline a flammable mixture may form inside the can (5-8.4).

5-9.8.3. *Filling Nozzles with Hold-Open Clip*

In some US states, gasoline dispensers known as "latch open devices" may be found that do not require operator pressure to hold the valve open at the nozzle. In these instances the operator may leave gasoline flowing to a tank without holding down the lever on the grounded nozzle. While ungrounded, the operator may develop a large static charge by walking on carpet in the station, getting in and out of the car or removing an item of clothing. Several fires have occurred due to a spark from the charged operator returning to the grounded nozzle. In one case the victim was seriously burned after withdrawing the nozzle in response to observing a small fire on the filling port, forgetting that gasoline was still flowing.

5-9.9. *Aerosol Spray Cans*

Propellants such as propane and butane are typically used in these cans; nonflammable halocarbons have been almost totally replaced due to the ozone layer depletion problem. The propellants are therefore usually flammable and may create flammable atmospheres when used carelessly in enclosed spaces. After six cans of insecticide had been sprayed a fatal explosion occurred in a ship's small galley after a spark from a refrigerator ignited the propane propellant [49]. Tests were conducted to explain a separate explosion which occurred during a can filling operation [49]. An equipment malfunction caused two cans to empty after the valves had been torn off. It was shown that only cans filled with powder + propellant charged sufficiently to cause a static ignition; cans filled with liquid + propellant produced only moderate charging. In this particular case, the cans had contained dry shampoo. It appeared that one or both cans, being isolated from ground by the plastic transport system, charged to about 35 kV and sparked to a nearby conductive object, perhaps another can. The can's 8.2 pF capacitance gave the maximum stored spark energy as 5 mJ, which will ignite propane-air mixtures over a wide range of flammable compositions. It was considered noncredible that an aerosol spray can could pose an ignition hazard when hand-held in normal use, although this could not be ruled out should highly resistive gloves be worn.

5-10. Cathode Ray Tube Video Display Screens

5-10.1. Cleaning

A static electric charge is commonly present on the face of cathode ray tube (CRT) screens, particularly on larger color monitors and color TV screens. This charge is the result of "writing" images on the phosphor-coated inside surface of the screen using an electron beam. The external electric field created by charge on the nonconductive screen can lead to brush discharges that may ignite flammable atmospheres. Also, if the screen is rendered partly conductive by application of conductive cleaning liquid while charge is still present, spark discharges capable of producing both ignition and unpleasant shocks can be produced. The application of cleaning fluids can in some cases supply both a flammable atmosphere and a means of producing energetic sparks. For example, commercial cleaners frequently contain conductive, flammable components such as isopropanol that are insufficiently diluted with water to prevent flammable vapor mixtures from forming. Also, some spray-on aerosol cleaners use a flammable gas propellant. It is recommended that CRT screens only be cleaned with the display screen switched off and after allowing time for the charge to dissipate. Where possible, nonflammable or high flash point combustible cleaners should be selected, especially if a charged CRT screen must be cleaned. Touch-screens may require the use of special cleaning agents. For hazardous locations, additional precautions are required (5-10.2).

5-10.2. Screens in Hazardous Locations

While some video display screens such as liquid crystal, gas plasma or vacuum fluorescent displays do not present the same charged screen hazards as CRTs, this does not imply that they are safe for use in hazardous locations. This requires special design and certification for use with a given flammable atmosphere. Non-certified equipment used in locations classified as hazardous under Article 500 of NFPA 70 "National Electrical Code" require a purged or pressurized enclosure to control ignition hazards as described in NFPA 496 "Standard for Purged and Pressurized Enclosures for Electrical Equipment." The screen in this case is located behind a window in the enclosure.

5-10.3. Static Dissipating Screen Overlays

Commercial conductive overlays may be used to drain CRT screen static in nonhazardous locations. The ground connection on such overlays must be made secure to prevent shocks.

6

POWDERS AND SOLIDS

Note: *"Powder" is a generic term for subdivided solid material, comprising pellets, granules and dust. Pellets have a diameter greater than 2 mm (U.S. No.10 Standard Sieve) and typically above 3 mm. Granules have a diameter between 0.42 and 2 mm, although granular powders typically contain dust. Dusts have a diameter less than 0.42 mm and as low as 1 μm. Suspensions of particles smaller than 1 μm are known as fumes.*

6-1. Flammability of Dust Suspensions

See NFPA 654 for general requirements. Combustible dust is defined in NFPA 654 as "any finely divided solid material 420 μm or smaller in diameter (material passing through a U.S. No. 40 Standard Sieve) that presents a fire or deflagration hazard." Excluding explosives and hybrid mixtures, it has been shown that this definition encompasses all powder suspensions that have produced deflagrations when tested according to ASTM E-1226 "Standard Test Method for Pressure and Rate of Pressure Rise for Combustible Dusts." Owing to the large increase of MIE, the static ignition hazards of powders are generally reduced by increasing the particle size (6-1.2).

Dust suspensions tend to accumulate increasingly finer material wherever there is an ongoing source of turbulence, such as when filling a storage container. This is because coarser particles entering the container settle out more rapidly, leading to an increasing concentration of fine dust in the suspension. In most cases, dust suspensions contain particles with a wide range of sizes and in order to assess flammability hazards, it is necessary to characterize the size of particles in different items of equipment. For example, during pneumatic conveying, powders tend to break up forming finer material which may accumulate either as a suspension or as a settled dust layer in bag houses or storage bins. Also, an additive added upstream at a low concentration may accumulate in the bulk dust fines; the probability of ignition downstream increases if the additive is more easily ignited than the bulk dust fines. Where the size distribution of particles is not known it should be assumed that suspended material is less than 75 μm (material passing through a U.S. No. 200 Standard Sieve). Samples sieved to sub-200 mesh are the largest normally tested for explosibility and minimum ignition energy (MIE). Where finer material is normally present in significant quantity or might accumulate, it might be necessary to take samples from equipment such as bag houses to estimate the "worst credible case" size distribution. Tests using sub-325 mesh or even sub-400 mesh may be required to assess explosibility and MIE. The following practicalities should be noted

a. *Soft Materials.* Some powders such as high molecular weight glycols and waxes may need to be frozen using dry ice or liquid nitrogen before they can be ground to a suitable size for testing. In some cases this problem can be avoided by obtaining samples from fines collection points in the production unit such as bag houses.
b. *Demixing.* Where mixtures of different powders need to be tested, precautions should be taken to ensure the relative concentration of the components remains the same in all size fractions tested. Otherwise, grinding and/or sieving may lead to concentration of individual components. An option is to grind and sieve the components separately then make up a mixture representing the size fraction and composition of interest.
c. *Residual Volatiles.* Samples are usually stored in a desiccator or in a sealed sample bottle under a suitable dry gas to remove moisture (see 6-1.6) and avoid degradation prior to testing. However, if the sample contains a volatile flammable component, such as the case with solvent-wet powders, the volatile material will diminish in storage. To evaluate such materials the techniques in 3-5.4.3 and 6-1.3.1 should be considered. If the flammable volatile concentration on a water free basis exceeds 0.2–0.5 wt% of the powder, the potential for hybrid mixture formation should be evaluated. If the concentration exceeds 1–2 wt%, it might be possible to exceed the vapor LFL, especially in hot, unventilated containers (6-1.3.2).
d. *Freshly Milled Metals.* The ignition energy of finely divided metals is unusually low when freshly milled, although it is difficult to conduct tests using typical MIE equipment, since dust surfaces rapidly oxidize on exposure to air. In some cases the reaction with air is so fast that

spontaneous combustion occurs. Special handling hazards are addressed in NFPA 651 (aluminum), NFPA 480 (magnesium), NFPA 481 (titanium), NFPA 482 (zirconium) and NFPA 485 (lithium). Several metals (Mg, Th, Ti, U, Zr) can burn in carbon dioxide, which is often used for inerting nonmetallic dusts, and metals such as Mg and Zr can even burn in nitrogen. Argon is sometimes used for extinguishing. Provided the limiting oxygen concentration (LOC) is greater than zero, such as aluminum, milling operations may be deliberately conducted in the presence of oxygen below the LOC so that surface oxidation takes place at a controllable rate.

6-1.1. Flammable Limits

Powders can be assessed as "nonignitable" if they meet any of the following criteria

a. *material is intrinsically noncombustible* such as silica, or will not burn in air when exposed to a temperature of 1500°F (815.5°C) for a period of 5 min (that is, NFPA 704 criterion for Flammability Rating of "0"). In rare cases the NFPA 704 criterion does not identify powders that can deflagrate in air under criterion (c), such as steel dust. However, the ignition energy of such exceptional powders is too high for them to be a "static" ignition hazard.
b. *material is handled so that >95wt% is larger than 40 mesh (425 μm)* and there is no potential for accumulation of finer material (unless the finer material is itself noncombustible, such as silicate talc).
c. *material will not deflagrate* when sub-200 mesh sample is tested according to ASTM E-1226

Powders should otherwise be considered ignitable/flammable/explosible. In the context of this book, these three terms are considered equivalent. If a flammable dust cloud may be formed during handling, or by suspension of an accumulated dust layer, tests may be required to characterize explosibility and ignition energy. There is often an advantage in conducting ignition energy tests first, since if the MIE is very high, as with carbon blacks, it might be concluded that the likelihood of ignition is negligible. Unless a design basis is required for deflagration venting, explosibility test data (6-1.1.3) might have no application. The combustion rate of dust clouds is faster for smaller particles. Explosibility tests (ASTM E-1226) should be performed on a sample representing the finest material that would credibly govern the overall combustion rate of the dust cloud. This depends on the application but might be considered as the sieve size through which the smallest 5 wt% of the dust passes. By subjecting a series of dust concentrations to a suitably energetic ignition source under standard test conditions, the following data can be found

6-1.1.2. *Minimum Explosible Concentration (MEC)*

This is the smallest concentration (mass/unit volume) that will support deflagration and is analogous to the lower flammable limit of gases. MEC values are typically in the range 15–60 gm/m^3. Minimum explosible concentration measurement should use the test apparatus described in ASTM E-1515 and not the Hartmann test apparatus. The latter is a less severe test and is likely to overestimate the MEC. Even when the ASTM E-1515 method is used, it is important not to overdrive the flame using a 10 kJ igniter in a 20-L test chamber. Either a 2.5 J igniter or a larger test chamber should preferably be used to avoid underestimating the MEC [184]. While in principle a "maximum explosible concentration" exists, this is not possible to measure accurately and has little practical application except perhaps to explain the lack of flame propagation through pipes during dense phase conveying. Values in the range 2000–6000 gm/m^3 have been reported.

6-1.1.3. *Maximum Pressure and Rate of Pressure Rise (P_{MAX} and K_{st})*

These explosibility parameters are used in assessing whether equipment will contain the maximum pressure developed during deflagration, or to design deflagration relief vents and other explosion prevention systems (see NFPA 68 and 69). The test method is given in ASTM E 1226.

If it is assumed that explosible dust will be present above the MEC, and equipment design data are not required, explosibility testing for P_{max} and K_{St} usually has no direct application. However, minimum ignition energy (MIE) testing should be considered to help determine the likelihood of ignition. Since MIE is extremely sensitive to particle size it is especially important to test a sample that is sufficiently fine to represent the worst credible case.

6-1.2. *Minimum Ignition Energy (MIE) of Dust Suspensions*

The maximum effective energy of a static discharge can be compared with the MIE of a dust to determine whether ignition is probable, as discussed in Chapters 2 and 3. MIE test methods are described in 3-5.4.2 and special considerations discussed in 6-1.3 through 6-1.5. At the time of writing, ASTM was close to development of a standard for dust MIE measurement.

Appendix A-6-1.2 shows that MIE increases with approximately the 2nd or 3rd power of representative particle diameter. This emphasizes the need to test samples that adequately represent worst credible case conditions. As previously discussed, samples are normally tested at sub-200 mesh and even finer samples may need to be tested according to the size of dust being handled. Owing to size classification during handling, the fine dust present in bag houses, on the walls of storage hoppers, and emitted from equipment open to the air may determine the representative particle size of dust that is in practice subjected to a static ignition source. The possible accumulation of ignition sensitive additives should be addressed; even though an additive is added at the 1000 ppm level, the fraction might be very significant in the bulk powder fines even if the correct concentration is added upstream of an accumulation point, such as a bag house. Human error can result in abnormally large additive additions.

6-1.2.1. Use of Narrow Particle Size Ranges

The flammability characteristics of powders depend on particle size distribution (PSD) with the most pronounced effect being on MIE (A-6-1.2). Particle shape also affects flammability owing to its influence on surface area. A variety of techniques are available to determine the PSD from which statistical quantities such as the surface or volume average particle diameter can be found; similarly, a particle shape factor may be assigned [197]. If sub-200 mesh samples are tested, this could represent any PSD with a maximum of 75 μm. Such a wide size range means that unless the powder is approximately monodispersed in the first place, such as *Lycopodium*, test samples are most unlikely to have similar PSDs. Consequently, MIEs determined for sub-200 mesh samples of the same powder can yield completely different results even in the same test apparatus. Also, MIE values obtained for a series of different materials might give an erroneous or possibly even an inverted hazard ranking. These problems can be minimized by testing samples having narrow size ranges. (See sieve size table in Appendix B, page 235.)

Minimum ignition energy hazards of different materials can be directly compared by collecting samples from the smaller of two sieves whose sizes differ by only a few microns. This eliminates most of the MIE variation due to size and shape. It also avoids disparities found in the literature due to the reporting of various statistical average diameters; if the PSD is very narrow, all of the commonly used averages converge to a single value. As an extension of this technique, the powder can be fractionated into a series of particle ranges by cascade sieving, enabling MIE to be found for a series of average particle sizes. Provided the sieves differ only incrementally in size from 200 down to (say) 400 mesh, there is no need to consider the PSD for each size fraction. For large capacity production especially, the MIE might be of interest for a wide range of particle sizes depending on the operation involved. The fractions must be carefully stored to avoid clumping and preferably tested as soon as possible.

In addition to the great dependence of MIE on particle size distribution and the possible accumulation of additives, complicating factors in measuring and applying MIE data include the presence of flammable gas (6-1.3.1) plus the effects of moisture (6-1.6) and possibly increased temperature (6-1.5) relative to the test temperature.

6-1.3. Hybrid Mixtures

Hybrid mixtures are mixtures of suspended flammable dust and flammable gas that together are flammable, although either the dust or gas component alone might be at a nonflammable concentration. Hybrid mixtures are especially hazardous since they combine the large charge densities of powder handling operations with the small ignition energy of most flammable gases. The effect of added flammable gas on the "static" ignition sensitivity of most dusts is usually small at gas concentrations less than 50% of the LFL [11]. However, in most practical situations, suspended dust containing flammable gas at a concentration exceeding about 10% of its LFL should be considered a hybrid mixture. This criterion allows for measurement errors, including the sensitivity of typical hand-held flammable gas analyzers.

6-1.3.1. Ignition Energy of Hybrid Mixtures

As a flammable gas is added to a dust suspension, the MIE usually decreases since gases usually have smaller MIEs than dusts. Direct measurement of MIEs for dust-gas hybrid mixtures is discussed in 3-5.4.3. However, since there is no standard test method and the composition of a hybrid mixture is difficult to define in many practical situations, it is often sufficient to estimate the value based on worst credible case assumptions. Figure 6-1.3.1 shows the hypothetical effect of adding diethyl ether to three different dust suspensions in air [11]. The general behavior is based on observations for dust–propane hybrid mixtures reviewed in [59]. The curve corresponds to the MIE of ether alone in air, whose lowest MIE value (LMIE) of 0.19 mJ is attained at an "optimum" concentration of 5.1 vol% ether. The sloped lines show the MIEs of hybrid mixtures of ether with three hypothetical dusts A–C whose MIEs in air are respectively 1000, 100, and 10 mJ. The MIE of each hybrid mixture (HMIE) decreases linearly on the semilogarithmic graph as ether is added. When the optimum concentration of ether has been added the HMIEs converge to the lowest minimum ignition energy (LMIE) of ether alone in air. This behavior can be modeled (A-6-1.3.1).

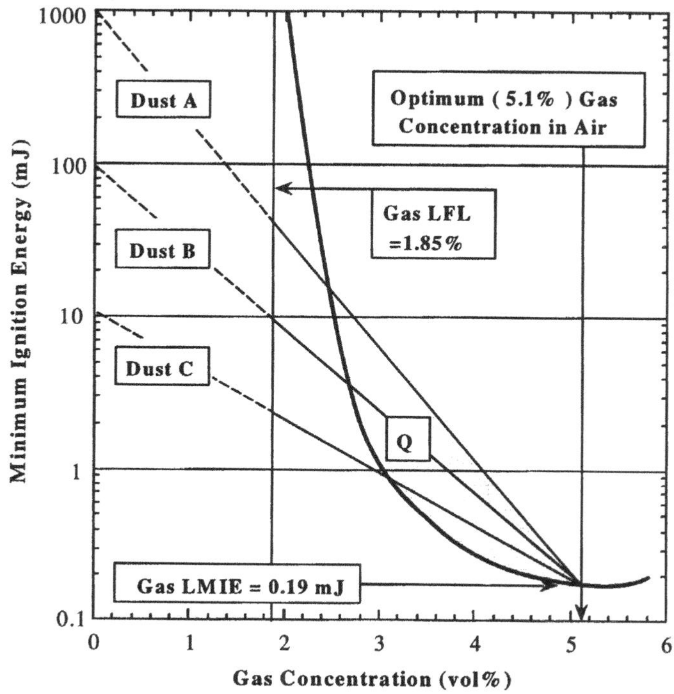

FIGURE 6-1.3.1. Hypothetical effect of adding ether to different dust–air mixtures.

The change of HMIE as flammable gas is added can be estimated using the HMIE equation [11]

$$\text{HMIE} = \exp[\ln D - (C/C_0) \ln(D/G)] \text{ mJ} \qquad (6\text{-}1.3.1)$$

where

D = MIE of dust alone in air (mJ)
G = LMIE of gas alone in air (mJ)
C = Gas concentration in hybrid mixture (vol%)
$C \leq$ LFL
C_0 = Gas concentration at LMIE composition (vol%)

This approach has not been tested for any dusts that burn heterogeneously (A-6-1.2), such as some metal dusts. The equation should not be applied for gas concentrations greater than the LFL [11] otherwise extrapolation might be made into region Q shown on Figure 6-1.3.1, where the predicted HMIE is greater than the gas MIE. The MIE of dust, D, must be determined by test using a conservatively fine dust sample to represent particles in the hybrid mixture. Values for G and C_0 can be found in Appendix B. Where G is not known it should be assumed to be 0.2 mJ unless sensitive gases such as hydrogen are present [11]. Where C_0 is not known or where a mixture of flammable gases is involved, it should be assumed equal to the stoichiometric concentration, which can readily be calculated for any gas mixture. The LFL of gas mixtures can be estimated using the Le Chatelier rule as described in NFPA 69.

6-1.3.2. Powders Evolving Flammable Gas

While it is relatively easy to remove volatile solvents such as ethyl ether and acetone during a drying process, less volatile solvents may persist at about 1 wt% or more in the product. The rate of evolution of solvents may be a slow process and difficult to estimate without practical testing. The worst case can be determined for a closed container at the highest anticipated temperature. For example, the temperature in a shipping container can be measured using a maximum–minimum thermometer so that the maximum transit temperature is found. This might exceed 50°C. Closed container evolution tests on a fresh powder sample can then be done to determine the maximum gas concentration that can develop. If this exceeds about 50% of the gas LFL, there is the potential for sensitized hybrid dust-gas mixtures. A separate problem would arise should the maximum gas concentration approach the LFL, which could indicate a gas ignition risk in the container. Tests have shown that powders containing upward of 0.2–0.5wt% solvent should be evaluated for these hazards. A possible solution, short of higher efficiency drying or extended powder storage in a ventilated area prior to shipment, is to use a high ventilation shipping container or ventilated hopper car. Special attention should be given to avoiding flammable gas evolution from powder packages that might be air shipped. A separate issue is powders that evolve flammable gas via ongoing reaction such as with atmospheric moisture (eg some metal alkoxides) or degradation of peroxides in the formulation. While the hybrid mixture hazards are similar to those with solvent-containing products, the source of flammable gas is in such cases renewable, and precautions need to be taken throughout the handling process. In all cases attention should be given to the possibility of gas concentrations markedly increasing with increased time and temperature. For example, a gas sample taken from a hopper car might underestimate the gas concentration that develops after shipment to a customer. Similarly, the gas concentration might increase after transfer to a powder silo heated in sunlight.

6-1.3.3. Powders Added to Flammable Liquids

See 5-4.5.1. Manual operations are particularly hazardous in terms of both frequency and consequence. If possible, direct introduction of powder to a vessel containing flammable liquid should be avoided by using an intermediate hopper plus rotary feed valve, or an equivalent arrangement isolating the flammable gas from the operator and powder package. Both the vessel and charging hopper can be inerted to achieve effective isolation; positive inert gas pressure on the hopper may reduce solvent-powder interaction problems. If this type of arrangement is not used, the vessel should where possible be inerted according to NFPA 69. This will not always prevent ignition where powder is introduced via an opening such as a manway, since air will be entrained with the powder while expelling flammable gas toward the operator. Where powder additions are from FIBCs, these problems become pronounced owing to the large volume of powder and high rate of transfer (6-7).

6-1.3.4. Drum Flakers

Numerous fires have occurred during flaking operations involving molten nonconductive material such as biphenyl poured onto rotating drums. As solidified material is flaked away by the doctor blade, the charged flake becomes separated from its counter-charge on the drum. Charge separation produces static discharges which may ignite flammable vapor or hybrid mixtures. The mechanism is similar to static discharges produced in web processes as the web separates from a roller. Since the molten solid may be as hot as 150–200°C a hybrid mixture may be formed by residual combustible solvent with a high flash point, or the vapor pressure of molten solid under the drum. Since the metal drum represents a grounded substrate the charge density of the solidified material may be unusually high prior to flaking and it is possible that ignition sensitive dusts might be vulnerable to ignition even with no significant vapor present. However, this is only possible if significant dust generation occurs. Ignition in flaker systems may also occur in the associated hopper collecting the charged, flaked material. Brush and bulking brush discharges in the hopper may cause ignition and the flame may subsequently propagate back to the flaker. This possibility should be evaluated before attributing the ignition to an event in the flaker itself.

6-1.4. Unstable or Energetic Powders

While this book does not cover shock-sensitive powders, such as primary explosives, UN-DOT Class 4.1 "Flammable Solids" are within its scope. These include thermally unstable powders that can both deflagrate in an oxidant and decompose in bulk. Examples include some nitrogen blowing agents. Should ignition occur at any point, a propagating decomposition front might travel through the bulk material. Because air is not needed to sustain decomposition, inerting and extinguishing techniques suitable for ordinary combustible powders may be ineffective. Also, because of the initially high packing density, very large pressures might rapidly be generated. Where such powders are handled in air, special attention should be paid to MIE testing and avoidance of static discharges capable of igniting the suspension and possibly initiating decomposition of the bulk powder. Tests should also be considered to evaluate ignition of powder layers and bulk material, with or without an air atmosphere.

Azodicarbonamide blowing agent ignited in the base of a conical orbiting screw mixer of volume 3.7 m^3 [163]. The "bottom initiated" confined decomposition led to a damaging explosion. The cause of ignition could not be determined with certainty. The possibility of a bulking brush discharge was not considered although it has been established that these may occur inside bulked heaps of powder to the container walls and floor (2-6.3.3). Initiation of bulk decomposition via a bulking brush discharge might be possible; in this case the mixer was heated by an 80°C water jacket and the powder had been vacuum dried, both of which would have decreased the powder MIE (6-1.5 and 6-1.6). However, the relatively small volume and inverted cone shape of the mixer are more supportive of a frictional initiating mechanism as advanced in [163].

6-1.5. Effect of Temperature on Ignition Energy

Dust MIE decreases with increased temperature, hence MIE values obtained at room temperature may not apply to operations conducted at elevated temperature, such as in dryers. Test work reviewed in [59] showed that the effect of temperature is more pronounced for dusts having larger MIEs. When MIE was measured at increasing temperatures for a series of dusts having different MIEs, a double logarithmic plot of MIE (mJ) against temperature (°C) yielded straight lines which appeared to converge at 0.088 mJ and 1000°C. Dusts with larger MIEs measured at room temperature therefore had steeper slopes on the log–log graph. While there is considerable room for error when generalizing this approach, a rough guide to temperature effects could be obtained by constructing a double-log plot of MIE versus temperature and extrapolating the MIE measured at room temperature to this convergence point (0.088 mJ and 1000°C). The MIE at any intermediate temperature could then be estimated by interpolation.

An analytical expression [unpublished] describing this behavior is derived in A-6-1.5

$$\text{MIE at temperature } T \text{ (Celsius)} = \exp\{\ln M - \ln(T/20) * \ln(M/0.088)/3.912\} \quad (6\text{-}1.5)$$

where

M = MIE measured at room temperature (assumed to be 20°C)

For example, a dust with MIE of 1000 mJ at 20°C has a predicted MIE of 21 mJ at T = 100°C. Based on the test results reviewed in [59] the effect of temperature on powder MIE might be major and should always be considered where powder is heated.

This approach has not been tested for any dusts that burn heterogeneously such as some metal dusts (A-6-1.2). Also, this large an effect of temperature has not been found by some other workers. It has yet to be confirmed that the effect is not due in part to loss of moisture (6-1.6) or evolution of flammable gas (6-1.3), both of which can have large effects on MIE.

6-1.6. Effect of Moisture on Ignition Energy

It has been shown that in the case of agricultural products (corn, tapioca, flour) at least, the MIE is increased significantly by increased moisture content of the dust [59]. The effect may be significant above about 1 wt% water, depending on the dust, and the dependence of MIE on water content may be comparable to its dependence on particle size (6-1.2). Where extraneous moisture might influence the MIE of a dust, the test should where possible be carried out on dried material. Where removal of moisture presents difficulties due to the nature of the material (for example, low melting point or thermal instability) the MIE test can be repeated after attempts to remove moisture have been made, such as extended storage in a desiccator or vacuum drying at a suitably low temperature. If the MIE decreases significantly, additional drying should be considered at least to the point where the material will normally exist under the conditions for which the MIE data will be applied. Moisture is not a problem with hydrophobic materials such as most plastic products, where storage in a desiccator prior to testing is all that is needed.

6-2. Charging Mechanisms

Powders can be charged by contact and separation between particles, or between particles and other surfaces such as bags or pipe walls. Upon contact, an electronic rearrangement occurs at the surfaces to minimize the free energy. Charge exchange occurs via electron transfer, based on the difference of effective work function, and, if ions are present, via ion transfer. The latter may occur, for example, due to the presence of water molecules at the surface. Separation of the surfaces after contact results in a net charge exchange, depending on the speed of separation. Charging may also occur when particles break apart or exchange matter with other surfaces. The process is enhanced by electric fields acting on the particles. In principle, even chemically identical particles may charge on contact, with the smaller particles tending to charge negatively. This is theoretically due to a decrease in work function with increased particle diameter [18]. In practice the surfaces of particles contain disparate chemical groups which affect the minimization of free energy and hence the charging that occurs.

When particles charge exclusively via contact with other particles, the change in net charge is zero since the suspension contains a mixture of positively and negatively charged particles whose total charge is the same as it was initially. This is complete bipolar charging. However, there is always a preponderance of one sign of charge on the coarser particles and an opposite sign of charge on finer particles. When powder is transferred into a large container the coarser particles settle out first carrying a net charge to the powder heap. The remaining suspension of fines carries an equal and opposite net charge. Both the powder heap and the suspension also contain a distribution of bipolar charge.

Charging between particles and boundary surfaces such as pipe walls creates net charge. A single sign of charge might be carried downstream by the particles (unipolar charging) while the countercharge remains on the boundary surface. If the boundary surface is conductive the countercharge flows to ground. If the boundary surface is nonconductive the countercharge accumulates, further modifying the charging process and possibly giving rise to energetic static discharges such as the propagating brush. Interparticle charging and particle-boundary charging usually occur simultaneously. The overall effect is therefore complicated and can change radically if some parameter is changed. Relevant parameters include materials of construction, humidity and flow conditions. Parameters affecting charging during powder conveying in pipes include flow velocity and mass flow density. Increased flow velocity usually increases charge-to-mass ratio by increasing the rate of particle contact-separation with pipe walls. Increased mass flow density usually decreases charge-to-mass ratio since the average collision rate between particles and pipe walls is decreased.

The charging process is affected by atmospheric moisture only to the extent that the moisture comes to equilibrium with the contacting surfaces.

The effects on suspended particles are complex. In a full-scale study it was observed that changing the relative humidity changed the net polarity of polyethylene particles being transferred to a silo [44]. The net charge at relative humidities less than 35% was positive, whereas above 35% humidity the net charge was negative. Above 50% humidity the magnitude of net negative charging current exceeded that of the net positive charging current observed at low humidity. This suggests that bipolar charging of polyethylene diminishes at high humidity in favor of negative charging. The significance of this with respect to static discharges on a compacting powder bed is discussed in 6-3.1.1.

6-2.1. Charge Density

The charge density of a powder can be described either on a volume or mass basis. The volume basis refers to the total volume occupied by the powder plus gas, whereas the mass basis refers to the mass of powder alone, so is equivalent to the charge-to-mass ratio. Typical charge-to-mass ratios [1] for nonconductive powders are given in Table 6-2.1, where these relate to the suspended powder prior to compaction. The table shows that a large range of charge densities is possible for a particular operation and values observed in practice will depend on both the particular powder and process used. Increased charge density corresponds to operations involving larger energy input and relative velocity between particles and equipment (6-3.1).

6-2.1.1. Effect of Particle Size on Charge Density

The maximum charge density on a flat, nonconductive surface is limited by the geometry with respect to conductive surfaces (C-2.5.3 through C-2.5.5)

TABLE 6-2.1. Typical Mass Charge Densities

Operation	Charge Density (nC/kg)
Sieving	0.01–1
Pouring	0.1–100
Scroll feed transfer	10 –1000
Grinding	100 –1000
Micronizing	100 –100,000
Pneumatic conveying	100 –1,000,000

and the breakdown strength of the surrounding medium. For the uniform field between a charged surface and a parallel grounded surface

$$Q_S = \varepsilon_r \varepsilon_o E \tag{6-2.1.1a}$$

Where in the case of atmospheric air, $E \sim 3$ MV/m at breakdown, hence

$$Q_S = 2.65 \times 10^{-5} \text{ C/m}^2 \tag{6-2.1.1b}$$

For particles, the maximum field intensity occurs at the particle surface and decays inversely with distance. This highly nonuniform field allows greater surface charge densities before breakdown takes place, depending on the curvature of the particle surface. Empirical studies reviewed in [18] give the relationship as

$$E = 9.3 \times 10^5 r^{-0.3} \text{ V/m} \tag{6-2.1.1c}$$

For $\varepsilon_r = 1$

$$Q_S = \varepsilon_r \varepsilon_0 E = 8.23 \times 10^{-6} r^{-0.3} \text{ C/m}^2 \tag{6-2.1.1d}$$

For spherical particles the total charge

$$Q_{TOT} = Q_S 4\pi r^2 \tag{6-2.1.1e}$$

The charge-to-mass ratio

$$Q/m = 2.47 \times 10^{-5}/(\rho r^{1.3}) \text{ C/kg} \tag{6-2.1.1f}$$

where r = density (kg/m³)

As an example, the maximum predicted charge-to-mass ratio of 20 micron spherical particles of density 1300 kg/m³ is 0.0244 C/kg or 2.4×10^7 nC/kg. This is 24 times the peak value given in Table 6-2.1 for pneumatic conveying. In practice, contact charging such as during powder flow will produce nonuniform charge and possibly bipolar charging, making the predictions increasingly conservative for smaller particle sizes. Also, this simple analysis neglects the field effects of surrounding particles. In any case, the maximum mass or volume charge density is larger for smaller particles. This usually results in a greater ignition hazard since not only is the maximum allowable charge-to-mass ratio increased, but the MIE of particles decreases rapidly as particle size decreases (6-1.2).

6-2.2. Classification of Powders Based on Conductivity

As in the case of liquids (5-2.5), the tendency of powders to lose charge can be simplified by classifying them into the three groups: (1) conductive, (2) semiconductive, and (3) nonconductive. The situation with powders is somewhat different from liquids because powders exist as heterogeneous mixtures with gas and their conductivities may be quite different from those of the pure solids; well-known examples are metal powders whose surfaces are oxidized. Powder conductivity depends on particle size, surface contamination and packing density. The conductivity of some powders can suddenly change once some threshold voltage has been applied that overcomes surface resistance between the particles. Hence it is found that some powders such as carbon blacks do not obey Ohm's law. Unless very low resistivities are measured at small voltages, such an effect might be present. When assessing the potential for electrical short-circuits and fires caused by resistance heating of powder, it is recommended that the voltage used exceeds the maximum source voltage in the electrical equipment involved. This plus attention to powder packing density will help mitigate surface resistance effects. When assessing static hazards due to semiconductive or nonconductive powders, the source voltage is typically 1–10 kV.

Demarcations between powders based on conductivity give only a guide to charge accumulation tendency. For example, dielectric constant is neglected even though Section 2-3 shows that relaxation time is proportional to this quantity. Other complicating factors such as impurities, surface conduction and effects of moisture are also neglected. The distinctions between the groups based on the type of test described in 3-5.1.1 become increasingly blurred at high volume resistivities owing to such extraneous effects.

6-2.2.1. Conductive Powders

These have a conductivity greater than 10^4 pS/m or a resistivity less than 10^8 Ω-m as determined using the type of test described in 3-5.6.1. Such powders, for example aluminum, become charged during flow but almost immediately lose this charge when transferred into grounded, conductive containers. Bulking brush discharges (6-4.2) do not occur on heaps of conductive powder. Provided the powder does not adhere, flow over nonconductive surfaces such as plastic bags might lead to brush discharges from the bag surface, which could ignite flammable vapor. If there is long term powder impact on a nonconductive surface, propagating brush discharges (6-4.3) might be produced; these have been observed on aircraft windshields following ice particle impact [167].

Charge may accumulate if conductive powder is transferred into a nonconductive container or a container with nonconductive lining. If charge accumulates, the powder may lose its charge in the form of a spark, allowing a large fraction of the stored energy to be dissipated instantaneously. For example, a fire might occur as a fiber drum with plastic liner is filled with a conductive powder. To dissipate charge a grounded metal rod is sometimes placed in nonconductive containers prior to the transfer. In one case, owing to an operational error, the rod was inserted when a plastic-lined fiber drum was partly filled with aluminum powder, leading to a spark from the powder surface to the rod which ignited the powder suspension. Although carbon blacks are conductive they are not a static ignition risk when handled in air owing to their very high MIE. Ignition would require the presence of a flammable gas at close to its LFL or the presence of volatile oils on the carbon black.

6-2.2.2. *Semiconductive Powders*

These have a conductivity between 10^2 and 10^4 pS/m or a resistivity between 10^8 and 10^{10} Ω-m as determined using the type of test described in 3-5.6.1. The special significance of powders in this range is that they are believed not to produce bulking brush discharges (6-4.2) while at the same time being too resistive to produce sparks. This is based in part on observations using sugar powder whose volume resistivity was in the semiconductive range [32]. Propane–air mixtures permeating from spherical probes lowered to the surface of sugar heaps in the test silo could not be ignited, although static discharges were detected electronically. This result suggests that only brush discharges were occurring. Bulking brush discharges or sparks terminating at the probe should have ignited the propane–air mixture (MIE ~ 1 mJ) despite some uncertainty in mixture composition (3-8.1.1). Pneumatic conveying of the sugar produced positive charging and brush discharges therefore originated from a positively charged powder surface. As discussed in 2-6.2 and reference [8], this polarity produces negative brush discharges whose effective energy is less than 1 mJ. This helps explain why no ignitions were produced.

6-2.2.3. *Nonconductive Powders*

These have a conductivity less than 10^2 pS/m or a resistivity greater than 10^{10} Ω-m as determined using the type of test described in 3-5.6.1. They do not give rise to spark discharges, but are more likely to produce other types of discharge (corona, brush, bulking brush and propagating brush) than powders in the other two groups. Owing to their high resistivity, these powders lose their charge at a slow rate and tend to accumulate charge even in properly grounded equipment such as powder silos. The degree of hazard due to charge accumulation rises significantly at resistivities above 10^{11} Ω-m. The least conductive powders, such as thermoplastic resins, have resistivities up to about 10^{16} Ω-m.

6-3. Pneumatic Conveying

This is the practice of conveying powders under positive gas pressure, which can lead to very high rates of charging (Table 6-2.1). The two modes are (1) dilute phase, which has lower powder/gas mass ratios in the range 3–9 and (2) more recently introduced dense phase conveying, which has higher powder/gas mass ratios in the range 10–60. Dilute phase uses higher gas velocities, typically 23–28 m/s, and lower pressures, typically 1–15 psig positive and 25–380 mmHg negative. Dense phase uses lower gas velocities, typically 5–10 m/s, and higher pressures, typically 10–50 psig positive and 240–600 mmHg negative [177]. Although the concept of "maximum explosible concentration" does not strictly apply, dense phase transfers may occur above the effective upper flammability limit of the powder while the dense phase transfer is in operation (6-1.1.2). Ignition hazards follow dispersion of the powder in equipment such as hoppers and dust collectors. Charge concentration as powder bulks in a container may lead to static discharges on the bed. If all metal equipment is bonded and grounded to eliminate sparks, "static" ignition sources are brush, bulking brush and propagating brush discharges as described in 2-6 and 6-4. An inert gas such as nitrogen is often used for pneumatic conveying where ignition sensitive dust suspensions are present (6-5). For general requirements see NFPA 654.

6-3.1. Charging in Pipeline Flow

General observations are that charge-to-mass ratio increases with increased flow velocity and decreases with increased mass flow density and increased particle size. This can be explained in terms of the collision frequency of particles with the wall, which produces net charge, versus interparticle collision which produces zero net charge. The number of particle–wall collisions per unit mass is increased as mass flow density is decreased. The charge transfer efficiency increases with velocity of separation, hence with flow velocity. Smaller particles can sustain a larger charge-to-mass ratio before ionization of the surrounding gas occurs (6-2.1.1). It follows that the largest charge-to-mass ratios are observed with small particles conveyed at high velocity and low mass flow density. This suggests dilute phase transfer. However, the charging current going to a receiving container is the product of charge-to-mass ratio and mass flow rate. High values of mass flow rate, typical of dense phase transfer, tend to reduce the charge-to-mass ratio by reducing particle-wall collision frequency. Therefore, the highest charging currents should occur at some intermediate mass flow density which might correspond either to dilute or dense phase transfer, depending on the conditions. There is no clear-cut demarcation between dilute and dense phase transfer in terms of mass flow rate and the velocity ratio between the two may be as small as about 2.

A complicating factor is that the inner pipe walls may become coated with the material being transferred and so change the nature of the charging process. Changes of polarity can occur. Plastic powders are notable examples. When plastic powders slide along pipe walls the frictional heat melts and deposits the plastic as a surface film. The film later breaks up into threads, commonly known as "streamers." In some cases the threads can lead to blockage. Studies with PVC showed that the charge density of the transferred powder decreased with the total mass of powder transferred, owing to continuous deposition of a surface film. Cleaning the pipe wall restored the original charge density with a high degree of reproducibility. The phenomenon could be avoided by roughening the wall surface to prevent sliding friction and in practice a threading machine was used to groove the walls [178]. The grooving technique might have application in avoiding plugging in small diameter pipe, but for large-scale systems, shot blasting is typically used to score the pipe surface and avoid streamer formation. The small scale studies showed that the cleaning procedure increased the charge-to-mass ratio by an order of magnitude [178]. In practical systems this phenomenon could be important during start-up of a new unit or after pipe cleaning since the cleaned pipe might increase the charging current or change the polarity of powder conveyed downstream.

6-3.1.1. Downstream Effects of Powder Charging

The ignition hazard in receiving containers depends on the ignition energy of the suspended powder and the effective energy of static discharges. The latter depends on the charge density and polarity generated on the powder during flow through the pipe. In addition to powder characteristics the polarity can depend on pipe material, surface deposits such as a plastic film and the humidity of the conveying air. A complicating effect is bipolar charging during pipeline flow. When this occurs the suspended particles in the container usually carry the opposite charge from heavier particles that predominate in the settled powder. The polarity may vary with position across the powder bed. This is due to various effects including local bipolar charging as powder impacts the heap and slides toward the wall, convection of smaller particles, and electrophoretic migration of charged fines in the ambient electric field. Reversals of polarity have been observed near the powder cone during periods of bulking brush formation [121].

During steady mass flow, low energy corona or brush discharges occur at high frequency on beds of nonconductive powder in receiving containers. Such discharges are incapable of igniting dust suspensions in air. Granular polyethylene was gravity-fed into an 11-ft.-diameter silo and the bed was video-photographed using an image intensifier. Corona or brush-type discharges occurred at the powder surface adjacent to the silo wall at very high frequency (>100 per minute). Discharge frequency maximized when the silo was 0.3–0.5 filled, at which time the discharges appeared randomly all over the bed surface [12]. Only a few bulking brush discharges were observed during several hours of observation. Ionization created by thousands of low energy discharges could be expected to suppress the formation of bulking brushes, which presumably requires transient accumulation of an unusually large charge density. It has been found that ionization is an important factor in neutralizing surface charge on powder beds. Polyethylene was artificially charged using a corona charge injection device and gravity-fed to a 0.5-m-diameter container. It was found that high negative charging currents resulted in smaller charge densities on the bed than were found at smaller charging currents. Owing to charge neutralization the probability of gas ignition was actually reduced when charging current was increased beyond a certain level [211]. This was demonstrated by bringing a gas-emitting probe to the surface of the charged bed. Ignition occurred most frequently for a moderate charging current and a high mass flow rate.

Since charge-to-mass ratios are typically larger for smaller particles (6-2.1.1 and 6-3.1) it follows that smaller particles should release charge via ionization at a faster rate as they bulk in a container. This helps explain why bulking brushes are more commonly observed during pellet transfers than during granular or dust transfers. Since pellet transfers are unlikely to generate flammable dust suspensions this explains why ignition is rare even though bulking brushes are commonplace in large, pelleted resin bins. Even if fine particles are present, mitigation is provided by bipolar charging effects. Large scale silo tests were made in which polyethylene dust was dispersed above positively charged pellets immediately before bulking brushes were anticipated. It was expected that this would constitute a realistic ignition test. However, the added polyethylene fines became oppositely charged and suppressed formation of bulking brushes [160]. As discussed later, ignition of polydispersed dust via bulking brushes might require predominantly negative charging so that charged fines have the same polarity as the coarse powder falling directly into the bed.

Consider first a dilute phase pneumatic transfer system operating at high velocity and relatively low mass flow density. As discussed in 6-3.1 this results in a high charging rate with minimal bipolar charging. Higher charge-to-mass ratios occur with smaller particles. As the powder bulks in the receiving container, excess charge is lost as soon as the electric field exceeds the breakdown field of air. If frequent discharges occur at widespread locations over the bed, the resulting ionization rate is sufficient to prevent large surface charge densities from accumulating. This prevents the generation of energetic discharges such as the bulking brush and limits the energy of brush-type discharges. Indeed, at very high charging rates the only significant discharge mechanism might be corona.

Consider instead an unsteady powder transfer operation in which slugs of highly charged powder are intermittently transferred at high mass flow rate to a receiver. The absence of an ionized layer during flow interruption could allow bulking brush discharge formation when powder flow resumes or increases. Bulking brush discharges are a recognized ignition hazard for some fine dusts in air (6-4.2.1). The ignition probability might be increased should unsteady transfer be accompanied by introduction of an easily ignitable additive, especially if operational error creates an unusually high concentration of suspended additive fines.

An "unsettled" question is whether bulking brush discharges are affected by bed polarity or might require large areas of oppositely charged powder to accumulate on the bed (2-6.3.5). As discussed in 6-2 it was observed in a full-scale test facility that the net polarity of high density polyethylene (HDPE) gradually changed from positive to negative as the relative humidity of the conveying air was increased [44]. At low humidity, bipolar charging occurred. The fines were charged negatively while the bulk of the material, determining the overall polarity, was charged positively. The polarity changed to net negative above about 35% relative humidity. It was concluded that charge build-up in HDPE can be minimized by operating at high mass flow rate and a relative humidity of about 35% [44]. This humidity corresponded to the observed null point at which the flow of positive charge was equal to the flow of negative charge and the net charging current to the silo was zero. However, if either predominantly negative charging or locally negative bed polarity is required for ignition via bulking brush discharges, it might have been concluded that HDPE transfers should be conducted at very low humidity so that charging is almost exclusively positive. As discussed above, bulking brushes can be suppressed by introduction of fine (easily ignitable) dust having opposite polarity from the bulk powder [160]. Therefore, one might expect fines to be ignited only if they carry the same sign of charge as the coarse powder, so that bulking brushes are not suppressed. For a given material, fine powder tends to charge negatively on contact with coarse powder. Hence, a possible criterion for ignition is that the powder is predominantly negatively charged. This has been observed with polyethylene at high humidity. There are anecdotal accounts of high humidity conditions existing before some polyethylene bin explosions; in one case a humidifier to "help control static" had reportedly been installed. There is no hard evidence for a generally applicable humidity effect with HDPE or even a generally applicable effect of polarity. The topic evidently requires more study. In general the best way to prevent ignition, short of inerting, is to control the ignition energy of the dust suspension by avoiding the accumulation of fine product and additive dust.

6-3.2. *Special Grounding Cases*

Normal bonding and grounding requirements are given in 4-1. Certain types of pipe connections may require jumper cables. The first general type is the grooved and shouldered joint and the second is the lap joint. In the first case, sections of pipe can become isolated by gaskets in the pipe connections. In the second, a flared or stub end pipe is connected via a bolted "slip flange." The latter fits loosely around the pipe and in many cases may be isolated from the pipe by paint or other coating. If a nonconductive gasket is used there may be no continuity across the connection. This can be corrected using a flexible graphite-filled gasket or other conductive gasket. Alternatively a jumper cable can be used. As shown in Figure 6-3.2, external sparks were observed from a carbon steel slip flange to an aluminum pipe conveying fumed silica through an electrically classified area. Both the flanges and pipe ends were coated with epoxy paint which effectively insulated them

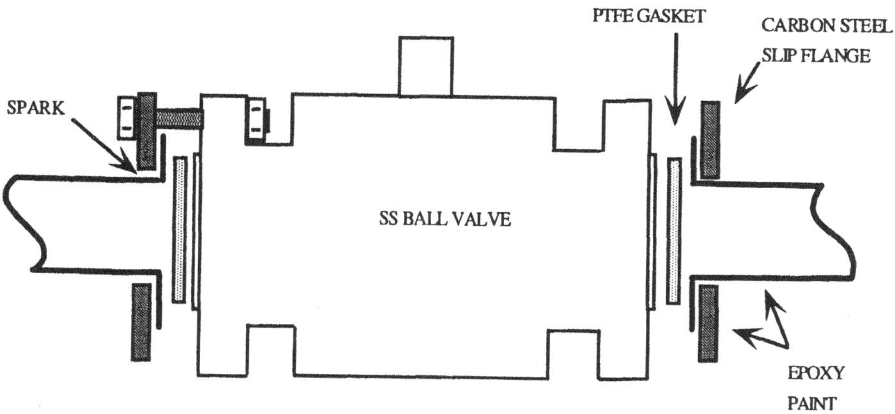

FIGURE 6-3.2. Ball valve isolated by twin slip flanges.

from each other. Although the slip flanges were electrically connected to the stainless steel ball valve via the bolts, the entire ball valve body was electrically isolated from the grounded pipe by PTFE gaskets plus the layers of epoxy paint. The situation was corrected by connecting jumper cables at each end between the valve body and pipe.

6-3.2.1. *Hoses*

These are discussed in 5-3.3 for liquid service, 5-4.6 with respect to vacuum trucks (liquid, powder or mixtures of both) and 6-4.1.3 concerning the possibility of sparks from unbonded spirals. Conductive and semiconductive hoses for powder service are available. The former include flexible metal types and the latter are available both with and without metal bonding elements; hence, it is always possible to specify a hose whose electrical continuity cannot be lost due to inferior construction or minor damage. Where there is no flammability hazard from the transferred powder, consideration should still be given to conductive/semiconductive hoses owing to the potential for puncturing of the fabric, operator shocks and static discharges outside the hose. External static discharges are potentially hazardous if the hose is located in an electrically classified area. Various considerations for hose selection include cost, weight, bend radius, conductivity, abrasion resistance, cold temperature tolerance, pressure rating and sensitivity to damage.

6-3.2.2. *Flexible Boots and Socks*

These are most frequently used in gravity transfer operations, where charging rates are relatively small compared with pneumatic transfers. Flexible boots are typically constructed of solid plastic or rubber while socks may be composed of finely knitted yarn. Boots are usually smooth-bored to minimize abrasion by the powder. A nonconductive boot may give rise to internal or external discharges including the propagating brush; the latter type of discharge cannot occur from knitted socks owing to the low breakdown strength introduced by the air gaps in the weave (2-6.2.4). As an example, an operator received a shock "like a spark plug" from a rubber boot on a flow diverter; the end-to-end resistance measured at 20 kV exceeded 10^{12} Ω. This was corrected by replacing the nonconductive boot with a semiconductive type, where the definition of "semiconductive" is the same as for a semiconductive hose. For flammable powders in air atmospheres, the end-to-end resistance of boots and socks should always be less than 100 MΩ, and preferably less than 1 MΩ as measured using a megohm meter with source voltage about 500 V or more. Should flammable vapor be present, the resistance should be less than 1 MΩ. Unless the service introduces severe cyclical stress on the flexible connection, such as in some sieving operations, it is easy to purchase materials yielding resistances less than 1 MΩ without compromising flexibility or service life. Semiconductive or conductive flexible connections should not be used to provide bonding between items of equipment such as fixed and portable metal containers. This is because in the event of breakage of the connection and loss of bonding, there is the potential for simultaneously generating a powder suspension and a source of ignition. Separate bonding or grounding should be used. Where convenient, flexible connections can be used with an external jumper cable for bonding. Note that an external winding of wire around a nonconductive flexible connection does nothing to dissipate static inside the connections, and can make matters worse (2-6.5).

6-3.2.3. Plastic Pipe/Hose Connections

Nonconductive plastic pipe should not be used for transfer of ignitable powder or for any powder transfer through electrically classified areas. Hazards comprise brush discharges, PBDs and external sparks due to induction. Charge accumulation is discussed in 5-3.2.1.

Short lengths of translucent, nonconductive plastic pipe or hose are sometimes used to visualize flow in otherwise conductive pipe or hose. They can give rise to all types of static discharge and should not be used for ignitable powders or in flammable vapor areas. Even in pellet service where ignition hazards have been ruled out, rigid high resistivity pipe may allow formation of PBDs which can give large shocks to personnel (2-6.5). Clear polyurethane hoses are available which are classed as "static dissipative" via Electronic Industry Association's Standard 541. Where these or equivalent antistatic hoses are used as flow visualizers, reinforcing wire spirals should be bonded at both ends. An "antistatic" hose might not be sufficiently conductive to prevent static accumulation under some powder flow conditions and its suitability must be carefully evaluated where easily ignitable powders are transferred or the hose is located in a flammable vapor area.

6-4. Types of Static Discharge in Powder Systems

Even where normal grounding practices (4-1) are observed, various static discharges (2-6) may occur in powder handling systems. Sparks can be avoided by providing a sufficiently low resistance from any conductive component to ground. Sparks nevertheless often occur due to oversight, design change, maintenance activities or mechanical failure. Charging rates in powder transfer are typically much greater than in liquid transfer and it is not always possible to eliminate brush and bulking brush discharges. Bulking brush discharges may ignite some fine dusts in air, provided the dust MIE is less than that of *Lycopodium* (<20 mJ). Brush discharges from isolated nonconductors such as plastic sheet or powder beds are only a hazard in the presence of flammable gas (6-1.3). Propagating brush discharges, which can ignite most flammable mixtures, can be avoided by not using plastic construction materials or by limiting the thickness where nonconductive coatings must be used (2-6.5). Dust might be ignited in air by transitional brush discharges from nonconductive layers such as plastic coated metal (2-6.2.5), but such discharges cannot readily be distinguished from PBDs. They represent an intermediate case between brush discharges and PBDs in which effective energy changes very rapidly with conditions.

6-4.1. Sparks

These may occur from

- conductive powder in a nonconductive container or container with nonconductive lining (6-2.2.1),
- ungrounded conductive components such as filter bag cages (6-4.1.2),
- poorly grounded conductors in proximity to a charged powder bed (6-4.1.1), or
- conductive objects inadvertently introduced, such as bag house components falling onto a charged powder bed.

6-4.1.1. Sparks Via Bulking Brush and Brush Discharges

Static discharges with peak currents measured in amperes might occur during silo filling, and the resistance to ground of any conductive object close to the powder bed should be sufficiently low to prevent sparking from the object to ground should it be struck by such a discharge.

In one scenario, deflagration of granular resin in a silo was considered most likely due to a bulking brush discharge on the powder bed striking an out-of-service capacitance probe, comprising a vertical cable (about 180 pF) attached to the wall by insulating spacers. The spacers provided multiple parallel spark gaps with a breakdown voltage about 15 kV [17]. The resulting spark ignited very fine dust which had accumulated on the cable and spacers plus suspended fine dust near the walls. The cable had an estimated resistance to ground of only about 40 kΩ, compared with the 1 MΩ normally considered adequate for static grounding. A dynamic analysis of cable potential was not possible owing to the unknown and variable rate of charging. However, charge transfers due to bulking brush discharges are known to be of the order 10 μC (2-6.3). If 1 μs is assumed for the effective discharge duration, the 10 ampere average charging current might have charged the cable sufficiently for spark breakdown to occur across the spacers. The scenario requires the effective duration of a bulking brush discharge to be comparable to, or less than, the RC time constant of the cable.

This hypothesis was incidentally supported by unrelated tests intended to show that bulking brush discharges can ignite dust suspensions; charge collected from bulking brushes inside a silo were diverted to a spark gap in a Hartmann Tube [161]. Dust ignitions were demonstrated despite a ground resistance of 1 MΩ in parallel with the spark gap. It follows that a ground resistance much less than 1 MΩ is necessary to prevent ignitions should the charge from a bulking brush be collected by a conductor near the powder surface. It is assumed that a parallel spark gap to ground exists and that the conductor has a sufficiently small capacitance to achieve the required potential for sparking.

Ignition of fine sulfur dust with a MIE of about 1 mJ or less was demonstrated by routing the collected charge (>280 nC) from a brush discharge through a spark gap in a Hartmann tube [34]. Unlike the previously described case for bulking brushes, there was negligible leakage resistance to ground parallel to the spark gap, hence the experiment does not provide any direct evidence that 1 MΩ is an excessive grounding resistance. As discussed in 2-5.1.1, brush discharges from isolated nonconductors typically transfer only 1–10% of the maximum charge possible via bulking brushes. If this ignition mechanism is possible at all, only very sensitive dusts or hybrid mixtures are likely at risk.

To avoid ignition via these indirect sparking mechanisms, the resistance to ground of conductive objects that could be encountered by discharges on powder beds should be less than 1 kΩ and preferably less than 100 Ω, so that the maximum potential attained is less than 1 kV.

6-4.1.2. Sparks from Filter Bag Cages

Sparks may occur from ungrounded cages to either the bag house structure or neighboring bag cages. There are two principal types of filter arrangement requiring slightly different grounding approaches

a. *Top Bag Removal.* Here the bag and cage assembly pulls out from the top of the tube sheet. A grounding strap is needed to ground the cage and venturi assembly to the tubesheet. The strap should be located in the plenum chamber above the tubesheet and comprises a ½-in.-wide stainless steel braided wire with eyelets at both ends.

After cleaning dust and deposits from the connections, one end of the strap is bolted to the cage/venturi assembly and the other to the tubesheet. The resistance between the cage and the tubesheet should be 10 Ω or less.

b. *Bottom Bag Removal*. Here the bags are generally made longer than the cages with the additional length being folded over the top of the cage to help secure the bag. The bag and cage assembly is then slipped over the venturi and a hose-type clamp is normally used to fasten the assembly to the collar. Owing to the folded bag, the use of nonconductive bag material will electrically isolate the cage from the tubesheet, and a dust deflagration due to this mechanism has been reported. A measure which has been successfully used involves nonconductive filter bags incorporating two short grounding straps at their tops, placed 180 degrees apart. These each comprise 24-in.-long × ½-in.-wide strips of stainless steel braided wire sewn into both the inside and outside surfaces of the bag, with 12-in. lengths sewn into each side of the cloth. The straps provide electrical continuity between the cage and the venturi collar when the bag is folded between them. Two straps are needed because some collars contain a slot. After installation, electrical continuity between the cage and tubesheet should be confirmed.

Plastic venturis should be avoided wherever there is the potential for isolation of the bag cage or other conductive component. Plastic venturi hazards are noted in [33]; a recent dust explosion originating in a bag house might have been due to their use. Criteria for selection of antistatic and conductive filter cloths are given in 6-5.2.1.

6-4.1.3. Sparks from Hoses

See 5-3.3 and 6-3.2.1. Powder transfer through hoses typically generates a greater rate of static charging than liquid flow, abrasion is greater, and the hose may frequently contain an ignitable mixture. Nonconductive hose incorporating an internal bonding spiral is not recommended except for pellet or noncombustible dust service in unclassified areas. This is because if the spiral breaks, internal and possibly external spark gaps may be created. In some cases, hose designs have omitted to bond the end connectors together via the spiral; sometimes there are two spirals present in the hose, only one of which is designed to provide bonding. Conductive hoses should preferably be flexible metal so that bonding does not rely on continuity of a breakable element. Other types of conductive hose should be designed so that end connectors are bonded to the grounding elements in the hose, while breakage of a conductive element cannot create a spark gap. To avoid spark gaps formed in this manner, a semiconductive hose design might be considered. Lightweight semiconductive hoses having no metal bonding elements are commercially available. These have a typical end-to-end resistance of a few thousand ohms per meter depending on hose size. In practice, various factors can be involved in hose selection (6-3.2.1).

6-4.2. Bulking Brush Discharges

See 2-6.3 and 6-3.1.1. These cannot always be eliminated in practical equipment above about 1 meter diameter where nonconductive powders are transferred, although flow rate reduction may be effective for the more conductive powders in this range (10^{10}–10^{11} Ω-m). Also, the use of active neutralizers has been proposed for relatively small silos (4-2.2.1). To minimize the probability of ignition, the first consideration should be whether the MIE of the powder being handled can be increased to greater than that of *Lycopodium* (>20 mJ), such as by minimizing the sub-200 mesh fines concentration and selecting additives having a MIE at least that of the sub-200 mesh product fines, particularly for containers that might accumulate additives. Where such methods are impractical, inerting is sometimes carried out, especially if powder properties make deflagration suppression and venting undesirable alternatives. Inerting is often considered where powder MIE falls below the 3–10 mJ range, according to the MIE test method used. Although these discharges would be expected to occur less frequently in smaller diameter silos, an increase of *L/D* ratio is usually undesirable for other reasons including its adverse effect on deflagration venting. Dust ignition via bulking brush discharge is uncommon but may explain the few silo explosions that occur where all metal components are properly grounded.

6-4.2.1. Effect of Particle Size on Bulking Brush Discharges

The maximum charge density of powders increases as particle size decreases, resulting in greater charging rates for equal mass flow rates (6-2.1.1). This in turn increases the frequency at which air breakdown occurs in the container and hence the frequency at which corona and brush discharges are produced. Frequent corona and brush discharges tend to produce an ionized gas layer above the bulked powder which dissipates charge from the surface region (6-3.1.1). Bulking brush discharges appear to be triggered by accumulation of some critical charge density in the powder heap, which is inhibited by the presence of the ionized layer. As a result, both the frequency and energy of bulking brushes in a given system tend to decrease as particle size decreases. Bulking brushes are most commonly observed during pellet transfer and it has often been stated that the systems most at risk are pneumatically conveyed coarse granules or pellets containing an easily ignitable fines fraction. Measured charge transfers have been found to increase both with particle size and silo diameter. A tentative model was developed for effective energy in terms of these two parameters [50]. As discussed in 2-6.3.4, a major application problem with the model is that the calculated effective energies are based on spark ignition tests and are unrealistically high. The values are similar to potential energy loss estimates made in 2-5.1.1, which greatly exceed the accepted maximum value of about 20 mJ deduced from dust ignition experience in industrial silos. The reason for the disparity is the low power density of bulking brush discharges relative to sparks in optimized spark gaps. The power density effect is more pronounced for dust than for gas ignition and it appears that the efficiency of nonspark discharges in igniting dust clouds may be only about 10% relative to gas mixtures of equal spark MIE (2-5.1.1). The effective energy of bulking brush discharges might be polarity dependent, the case being made in 2-6.3.5 that ignition might require predominantly negative charging of both the fines and the bulk powder. Although coarse powder tends to give a greater frequency of bulking brushes, the presence of easily ignitable fines does not by itself give rise to an ignition hazard. This is because if bipolar charging occurs the fines tend to neutralize the bed and suppress bulking brushes (2-6.3.5). In a typical industrial situation involving powder having a wide and nonconstant size distribution it is useful to consider the maximum effective energy of bulking brushes to be about 20 mJ in large silos. The value decreases with decreased charge transfer and hence with decreased container diameter [see Eq. (2-6.3.2)].

6-4.3. *Propagating Brush Discharge*

See 2-6.5. PBDs can be avoided by simply not using nonconductive equipment such as hoses, wands, pipes, flexible connections and linings. The phenomenon does not occur where insulating layers are due only to loose, nonconductive powder accumulations, since gas spaces between the particles lower the breakdown strength relative to a continuous nonconductive layer. Hence, PBDs do not occur due to loose powder layers on silo walls and other surfaces. Criteria for avoiding PBDs from nonconductive layers are given in 2-6.5. Linings can be made sufficiently thin so that they cannot sustain the minimum surface charge density required; electrical breakdown results in weaker transitional brush discharges described in 2-6.2.3 and

2-6.2.5. These weaker discharges may nevertheless ignite some dusts in air since their effective energies include the range between ordinary brush discharges (5–10 mJ) and PBDs (100–1000 mJ).

Painted coatings having a breakdown strength of about 400 V/mil should usually be too thin to be a problem; painted coatings are typically less than 7 mils in thickness and electrical breakdown should therefore occur at less than 4 kV (2-6.2.4 and 5-4.1.3). However, where the breakdown strength is uncertain or exceptionally thick layers might have accumulated owing to repainting, the situation should be evaluated using thickness gauges and suitable test coupons. For slow charging rates the 4 kV criterion should normally be conservative (2-6.2.1). However, in containers such as silos and hopper cars very large surface charge densities can be almost instantly deposited on insulating wall coatings by bulking brush discharges. Although this mechanism has not been studied in detail, it should be assumed that such conditions allow little or no margin of safety when the 4 kV criterion is applied.

6-5. General Operations

Unlike the case with liquid operations, where numerous restrictions are applied for safe handling (particularly for nonconductive liquids), powder operations are normally considered only with respect to the provision of proper grounding unless MIEs less than about 10 mJ are involved or flammable vapor is additionally present. For powders in air

- Powder transfer velocity and mass flow rate are not subject to any maximum values,
- Slow starts are not used,
- Bottom filling is not required and is normally impractical and,
- Inerting is rarely used for powder storage, although its use is increasing for pneumatic conveying, especially for ignition sensitive dusts.

For most powders, static discharges are *expected* to occur when powders are transferred in bulk, even under gravity filling conditions. The reasons that static discharges are normally tolerable are

- Corona and brush discharges from compacted beds will not ignite dusts in air.
- Sparks, transitional brushes and PBDs can be prevented.
- Bulking brushes have a very small ignition probability in most cases (as discussed below).

- Deflagration venting (see NFPA 68) is highly applicable to powder operations. This is because most powder operations are at atmospheric pressure, the rate of deflagration pressure rise is usually small enough for vent relief panels to be of a practical size, and the subsequent fire is limited because most powders cannot burn in bulk.

If proper bonding and grounding are employed, there are no nonconductive plastic materials of construction, and hybrid mixtures avoided, static ignitions can only occur via bulking brush discharges during container filling or other operation that results in charge concentration. Some texts at this point mention the avoidance of "tramp metal," but unless the objects involved are large enough to have significant capacitance they should be considered mechanical spark or friction hazards rather than static hazards.

As discussed in 2-6.3 and 6-4.2, powder ignition via bulking brush discharge requires the coincidence of several factors, each of which has an associated probability. The minimum conditions are

1. Discharge must coincide with flammable powder suspension in time and space,
2. Discharge must have effective energy sufficient to ignite powder, i.e.,
   - 2a. hottest part of discharge (at container wall) must coincide with optimum powder concentration,
   - 2b. local powder suspension must comprise small particles with low MIE and,
3. Local ignition must be followed by flame propagation into surrounding suspension above powder MEC.

If a powder has been characterized by a particular MIE, say 10 mJ, this does not mean a typical suspension in air has a MIE of 10 mJ. MIE tests are made on either sub-200 mesh dust or finer fractions designed to represent the worst credible case (6-1.2). The reported MIE corresponds to the optimum concentration of this fine dust fraction in air, which is relatively large and often difficult to achieve in practical operations, being typically in the range 250–750 g/m^3. Estimates of the effective energy of bulking brush discharges, based on accident analyses, indicate a maximum value of 10–20 mJ, or less than the MIE of *Lycopodium* (20 mJ). Since the effective energy exists as a distribution of possible values up to the maximum value, most discharges will have only a fraction of the maximum effective energy. The probability of ignition is extremely small for dusts with MIE similar to *Lycopodium* and negligible for dusts with significantly greater MIEs. Since the ignition frequency for most dusts is very small, it is common practice to provide only deflagration venting for containers and other equipment operating at low pressure. Additional or alternative measures such as inerting, containment or flame suppression should be considered depending on the risk factors involved. Such factors include the ignition sensitivity and toxicity of the dust.

Although the charging currents produced in most flammable liquid handling operations are less than those produced during powder conveying, the ignition probability for many flammable liquids handled in air is much greater than for a flammable dust suspension. This is related to the ease with which a uniform vapor–air mixture can form throughout an enclosed volume. If the handling temperature allows formation of the most easily ignitable vapor–air mixture, this mixture can persist throughout the volume for a large fraction of the filling time and can be vulnerable to ignition even by relatively weak positive brush discharges.

The high charging currents and long relaxation times associated with many powder transfers can create a very high frequency of brush discharges during container filling, superimposed on which may be less frequent bulking brush discharges. Should a flammable gas be present in the container, the probability of ignition increases rapidly as the gas concentration increases from about 50% of its LFL toward its optimum concentration, which is usually slightly rich of stoichiometric. It is essential to purge flammable gases from resins such as polyethylene before transfer is made to a non-inerted storage silo.

Sections 6-1 through 6-4 can be applied to most powder handling processes. Where powders have small MIEs compared with *Lycopodium* and will be bulked into a container, inerting may be considered. Inerting is often used where powder MIE is below 3–10 mJ, depending on various considerations including MIE test method, practical experience with the powder, availability of inerting gas and consequences of an ignition.

6-5.1. Vacuum Trucks

In early 1998, an explosion and fatality occurred while vacuuming off-specification granular polypropylene via the side manway of a bin to a high capacity vacuum truck. The hose was 6-in.-diameter nonconductive plastic with a metal lance that was inserted through the manway. The flame did not propagate through the hose to the truck and equipment damage was confined to the bin and dust collector. It is believed that a spark discharge occurred from the ungrounded lance to the bin but unclear as to whether the bin contained any flammable vapor; initial reports were that the resin had been properly purged. If the presence of fine polypropylene powder is assumed, it is not unreasonable that spark ignition could have taken place in the absence of flammable vapor since the powder MIE in air might have been less than 20 mJ. From available information this incident should have been avoidable by using a conductive or semiconductive hose bonding the lance to the truck as discussed in 5-4.6 under "vacuum trucks" and elsewhere under "hoses."

6-5.2. *Bag Houses*

Grounding of cages is discussed in 6-4.1.2. In addition to providing proper bonding and grounding, the probability of ignition can in some cases be reduced by decreasing the air-to-cloth ratio in dust collectors (that is, the air flow rate divided by the total effective filter area). Higher air-to-cloth ratios tend to classify the finer particles in the top part of bins and dust collectors; owing to the relatively high frequency of maintenance activities in bag houses, the potential for grounding error and spark ignition should be considered. Typical air-to-cloth ratios are in the range 5–7 ft/min for granular materials and 3–5 ft./min. for fine (sub-200 mesh) powders.

6-5.2.1. *Filter Cloth*

Except where conductive dusts are collected, the only credible type of discharge from a charged filter cloth is the brush discharge (2-6.2.1). These cannot ignite dusts in air atmospheres so conductive or antistatic filter bags are not needed. Instead such bags might create spark hazards should parts of the fabric become electrically isolated or should a bag filter fall into a bin. Several dust explosions have been blamed on loss of electrical continuity in antistatic filter cloth. These involved bag filters or rectangular filter banks with various types of antistatic filter fabric.

Conductive dusts might be collected nonuniformly and in principle could give spark discharges from isolated patches of dust. Carbon blacks typically have very high MIEs and are not subject to static ignition, whereas metal dusts can have very low MIEs and might be ignited by this mechanism. Also, in the presence of flammable gas, hybrid dust–gas mixtures might be ignited by brush discharges. If metal dusts are collected or if hybrid mixtures cannot be discounted, antistatic or conductive filter cloths should be selected. For ignition sensitive metal dusts such as magnesium, completely conductive fabric should be selected and all parts of the bag house should be very carefully assessed for grounding. Of the available types of "antistatic" and "conductive" filter cloths, only some are effective. For example, a bag filter might simply contain an embedded loop of wire braid which will do nothing to dissipate static charge from collected dust on the cloth. Antistatic cloths containing loose metal fibers have the disadvantage of randomness, which might allow sparking from isolated conductive patches. Cloths that are uniformly treated with antistatic agents avoid this problem, but unless the antistat is highly visible there is the potential for manufacturing or installation errors. Ideally, the cloth should be manufactured entirely from semiconductive or conductive fibers and installed so that grounding cannot be lost.

6-6. Manual Transfers from Portable Containers

Where flammable gas/vapor is present see 5-4.5. For powder transfers in air it is necessary only to avoid sparks by proper bonding and grounding of all conductive components such as funnels and the metal chimes on fiber drums. Grounding of people (4-3) should be considered for fine dusts whose MIE is less than that of *Lycopodium*. Paper bags are unlikely to produce spark hazards although this is theoretically possible from wet patches or large staples, etc. A special case is where either paper or plastic bags contain a metal layer applied either as a surface film or as an inner-plied foil layer. Such layers are sometimes used as moisture barriers. The conductive metal layer can become charged during powder flow and create a spark ignition hazard unless special grounding measures are taken; in 1999 a dust ignition incident was reported involving a plastic bag with an inner plied aluminum foil layer. Except at relative humidity less than 10–20%, paper bags and chimeless fiber drums can be considered semiconductive and may be grounded by contact with grounded equipment if a bonding clip is not used. Alternatively, they can be grounded via a grounded operator provided nonconductive gloves are not worn. A special case of portable container transfers is the FIBC (6-7).

6-7. Flexible Intermediate Bulk Containers (FIBCs)

For a comprehensive description of hazards including accounts of fires and explosions see [10]. FIBCs are collapsible rectilinear plastic containers for solids handling, comprising a body constructed of film, woven fabric or other flexible material plus any appropriate service equipment and handling devices. Typical volumes are not more than 1.5 m^3 and capacities 400–1000 kg. Filling and discharge are typically done through spouted top and bottom openings, where the bottom spout requires only unfolding/untying to begin gravity unloading. However, some FIBCs feature fully open tops or bottoms that require cutting of the fabric. FIBCs can also be vacuum unloaded (6-7.4). During filling, ignition hazards of the powder may be mitigated by inflating the collapsed bag with nitrogen. This is done primarily for product quality reasons and there are no available reports of ignition during FIBC filling. When emptying a FIBC at uncontrolled rate, inerting is impractical since air will be rapidly inhaled as the FIBC empties; complete emptying might take 10–30 s. During uncontrolled emptying into an atmosphere containing flammable vapor, such as at the port of a mixing vessel, the large flow of air entrained with the powder will counteract any inerting in the vessel while expelling flammable vapor toward the operator. If the FIBC is not properly vented at the top, flammable vapor might be inhaled into the FIBC and ignite. If a loose liner is used, it should be designed so that it does not leave the FIBC and present an ignition risk (5-9.5). Owing to the large flow rate of powder and its associated generation of static electricity, potential for charging of ungrounded operators, plus air entrainment effects, uncontrolled FIBC unloading in flammable vapor atmospheres is intrinsically more hazardous than transfers from 25 kg bags (5-4.5.1).

A proposed European FIBC classification scheme [152] referring to Type A, B and C containers is used in the following sections. Type D containers have been added to reflect recent developments. In brief, these FIBCs are

Type A. General 100% plastic types that cannot be grounded.
Type B. Thin-walled 100% plastic types that cannot be grounded.
Type C. Conductive types that require grounding and,
Type D. Antistatic types that do not require grounding.

There is no available evidence that PBDs occur in FIBC operations and therefore no demonstrated need for thin walled Type B containers designed with a breakdown voltage of 4 kV or less [10]. However, since Type B containers are commercially available and in principle may offer slightly less risk, they might be considered where thin walls will not have a detrimental effect on product quality (6-7.1.1).

6-7.1. Powder Transfers in Air Atmospheres

Provided the powder does not evolve flammable vapor, the FIBC selected should be of a type which does not require grounding, since FIBC manufacturing defect or operator error that results in loss of grounding will introduce an ignition hazard due to spark discharges. Types A or B can be used for nonfrangible pellets without regard to grounding, since there will be no flammable dust suspension present. Where significant dust is present, an ungrounded Type A or B can be used provided the operator is grounded (4-3). At greater cost, a Type D could alternatively be used in either case. If Type C is used there will not only be additional cost but also a spark ignition risk if proper grounding is not carried out.

Where an easily ignitable dust is emptied from Type A into equipment, the spout should preferably not be introduced through a metal nozzle so as to leave only a small clearance between the plastic spout and the underlying metal of the nozzle. This is because of possible generation of transitional brush discharges on the spout (2-6.2.3 and 2-6.2.5). Also, where ignition sensitive dusts are unloaded in air the spout should preferably not rest on a metal surface. In the absence of an air gap the underlying metal surface increases the maximum charge density that can accumulate on the inner surface of the spout and this might result in an incendive discharge. Where such conditions cannot be avoided a Type B might be considered (6-7.1.1).

If a powder is capable of evolving flammable vapor at or above its LFL it should be considered as being handled in a flammable vapor atmosphere (6-7.3). This hazard may be present if the powder is reactive (slowly decomposes or reacts with moisture forming flammable vapor) or contains upward of 0.2 wt% of flammable solvent. Evaluation of a vapor hazard may require closed container testing over an extended period at a sufficiently high temperature to represent worst-case storage conditions. This may be several weeks at 50–60°C in some cases. If the vapor concentration can attain greater than 50% of its LFL the combination of vapor plus suspended dust may result in an easily ignitable hybrid mixture which should be considered equivalent to a flammable vapor atmosphere. Types A and B should not be used in flammable vapor atmospheres. For direct transfers to flammable liquids see 6-7.2.

6-7.1.1. Low Breakdown Strength Plastic FIBCs

It has been hypothesized that propagating brush discharges (PBDs) might be produced when either filling or emptying a 100% plastic FIBC. Unlike other discharges from plastic surfaces, PBDs may ignite dust suspensions in air. Also, as discussed in this book, less energetic transitional brush discharges might ignite some dusts under conditions that do not support PBD generation. Thin walled Type B containers are commercially available with a wall breakdown voltage of about 4 kV or less. This design criterion avoids both types of discharge (2-6.2.3 and 2-6.5). However, the lack of either clear evidence or an experimental demonstration of these phenomena in FIBC operations suggests that they have a low probability and might not occur in practice [10]. If a Type B is used, adverse effects may include increased moisture ingress and vulnerability to splinters from pallets. This can have a detrimental effect on product quality, particularly for hygroscopic solids. Given the absence of any reported FIBC loading fires, it was suggested that an alternative to thin walled FIBCs was simply to perforate the spout [10]. This would drastically reduce the breakdown voltage, preventing PBDs from occurring in the region of greatest powder-wall charging while minimizing quality problems due to thin walled FIBCs. Since there is no specific guidance on the need for thin walled FIBCs, this should be a risk based decision. If the risk of Type A is seen to be unacceptable, for example if an ignition sensitive or toxic dust is being handled, and Type B cannot be used owing to moisture sensitivity, Types C or D should be considered instead (6-7.3).

6-7.2. Powder Transfers from FIBCs to Flammable Liquids

This situation is distinct from FIBC use in a Class I Div. 2 electrically classified area. The receiving container typically contains flammable concentrations of vapor. Unless a noncombustible powder or nonfrangible pellet is being handled a hybrid mixture will form in the receiving container and the safety factor normally applied to liquid flash point (5-1.1.5) does not apply. Flammable vapor at only a fraction of its LFL may significantly increase the ignition sensitivity of a flammable dust suspension. Transfers should preferably be done via an intermediate grounded metal hopper with rotary valve or its equivalent, keeping both the mixing vessel and the hopper inerted throughout the transfer. The vessel vent line should be appropriately sized to minimize flammable vapor expulsion into the hopper. These precautions isolate the flammable vapor from the FIBC and operator. While this method reduces hazards due to the FIBC design, the transfer will still be made in the potentially flammable vapor atmosphere surrounding the vessel so Types A and B should not be used. Instead the FIBC should be either Type C or D (6-7.3). If transfers are made directly from a FIBC to a flammable liquid without an intermediate purged hopper, not only must superior attention be given to assuring proper vessel inerting but a residual risk must be assumed whatever the FIBC type. This risk is due to the interface of flammable vapor and air, at which various ignition sources may occur during the transfer (5-4.5.1).

6-7.3. Conductive and Antistatic FIBCs

These are designed primarily for use in flammable vapor atmospheres and should also be considered where powders evolve flammable vapor as described in 6-7.1. Various designs of conductive Type C, and antistatic Type D, FIBCs are commercially available. Whichever type is selected for use in flammable vapor atmospheres, the FIBC manufacturer should provide both quality assurance and test data showing that full scale ignition testing has been successfully carried out. Compatibility, strength, and economic considerations may in practice restrict the choice of available FIBC designs.

Type C typically contains a conductive grid with mesh spacing of about 20 mm, having a maximum resistance to ground of 100 MΩ from any point contacted by a 25 mm test electrode. All types require grounding, usually via at least one grounding tab. To offset operator grounding error, some types are designed so that the slings are conductive, hence the FIBC will be automatically grounded via the slings provided the hoist system provides a ground path. Advantages of Type C designs may include compatibility and re-use capability. In addition to the great likelihood of hazardous spark discharges should proper grounding not be present, disadvantages may include cost and difficulty in recycling due to metal fibers.

Type D does not meet "conductive" specifications but is nevertheless sufficiently conductive to dissipate charge, as demonstrated by full scale testing. The design typically incorporates a system of independent conductive fibers whose capacitance is too small to give incendive sparks at the maximum voltage attained; the latter is limited by corona discharge above 1–2 kV plus the action of a topical antistatic agent applied to the fabric. Type D does not require grounding and therefore avoids operator grounding errors. While this feature is a major advantage over Type C, compatibility of the product with the antistatic agent should be considered.

During filling or emptying, Type D produces a substantial transient electric field which may induce hazardous charges onto ungrounded objects [60], making it especially important that personnel and conductive equipment should be grounded. It was reported that during filling of a Type D with charged polypropylene pellets it was easy to produce incendive sparks from a paper clip attached to the FIBC. This was demonstrated using a propane mixture having a MIE of about 1 mJ. Ignitions during filling were also obtained via sparks from a conductive wet patch on the FIBC [226]. The far more important practical case involves emptying of an initially uncharged powder. Although ignition via the "wet patch" mechanism has been demonstrated during the emptying of polypropylene pellets, it has yet to be shown that this can occur when the powder is initially uncharged. This represents the practical case since powders typically reside in FIBCs for many days or weeks before the FIBC is emptied. Although operators and equipment can be grounded, it is difficult to rule out the presence of a small conductive object or wet patch on the floor or spout of the FIBC. If a Type D is used for direct powder transfer to flammable liquid it is important that this scenario be ruled out by realistic testing.

6-7.3.1. Ignition Testing for Conductive and Antistatic FIBCs

FIBC testing has been carried out using a grounded metal test probe held in a tube containing a flammable hydrocarbon-air test mixture (typically ethane or propane in air). The probe arrangement is typically brought to the external surfaces of the FIBC during filling or emptying operations. However, since powder transfer might be made to methanol which has about half the MIE of propane (Appendix B), it is recommended that ethylene at its most easily ignitable composition be used to simulate flammable vapor atmospheres rather than aliphatic hydrocarbons. Since ethylene has a very small MIE relative to common solvents, this offsets limitations in the number of tests that can reasonably be done and the probability that worst case conditions will not be realized under test conditions (3-8.1).

6-7.4. Vacuum FIBC Transfers

See 5-4.6, 5-8.5 and 6-5.1. Hoses should be antistatic or conductive with special attention paid to grounding of metal connectors. Even if noncombustible powders or combustible pellets are handled, nonconductive hoses may give rise to discharges capable of igniting flammable vapors and creating nuisance shocks to ungrounded operators. Consideration should be given to personnel grounding systems, especially where easily ignitable dusts are being transferred.

APPENDIX

A

EXPLANATORY MATERIAL

A-2-6.5. Propagating Brush Discharge

The mechanism of surface flashover in a PBD may be visualized with reference to the circuit shown in Figure A-2-6.5. A 1-μF capacitor "C" is charged to 20 kV through a large-value resistor "R" which isolates the power supply. A source voltage of 20 kV will normally produce sparks about 6 mm long between simple electrodes in air. However, given an extended system of spark gaps and initially uncharged capacitors as partly shown, sparks can propagate 50–100 times as far owing to the very steep gradient produced at the stepped discharge wavefront. A typical vertical column arrangement can produce sparks propagating along the electrode system to about 0.5 m [176]. In a PBD situation, a complex system of capacitance exists between and among charges on the surface and their corresponding countercharges on the opposite side of the charged layer. The resulting system of virtual capacitors, across and through the charged layer, is analogous to the circuit shown in Figure A-2-6.5. A propagating "step by step" discharge spreads laterally

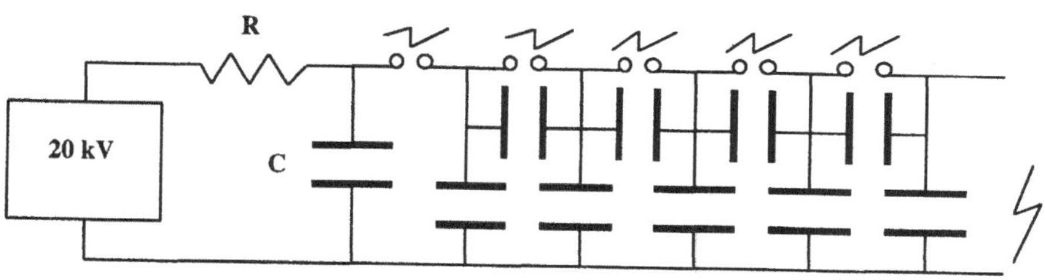

FIGURE A-2-6.5. Step by Step discharge mechanism producing long discharge with relatively low voltage.

over the charged surface away from an initiation point once flashover conditions are attained.

Similar considerations may be applied to bulking brush (2-6.3) and surface streamer (2-6.6) discharges which propagate across charged surfaces over distances up to about 1 m at source voltages less than 100 kV.

A-4-1.3. Resistance to Ground

The resistance to ground should be sufficiently low so that neither the MIE nor the minimum ignition voltage can be attained at the maximum anticipated charging current to the system. The resistance required depends not only on the flammable mixture but also on the system being charged. For simple capacitance sparks, the MIE of gases is decreased as storage capacitance and electrode tip diameter are decreased. Since gases have smaller MIEs than dusts, they represent the more challenging case for grounding. The required resistance to ground can be found by considering the worst case flammable mixture, storage capacitance and electrode diameter.

Table A-4-1.3a, based on data in [62], shows how the MIE (mJ) of 28 vol% hydrogen and 8.5 vol% methane in air vary with circuit capacitance (pF) and electrode diameter (mm). "Points" refers to the use of steel gramophone needles. The table shows that MIE is decreased with decreased capacitance and electrode diameter; however, as reflected in Figure 3-5.4.1 corona

TABLE A-4-1.3a. Minimum Ignition Energy (mJ) Variation with Capacitance and Electrode Diameter

	Capacitance (pF)	15 mm diameter	1.59 mm diameter	0.5 mm diameter	Points
Hydrogen	146	0.390	0.260	0.210	0.073
	50	—	0.160	—	—
	30	0.094	0.079	—	0.034
	6.1	0.022	0.016 = LMIE	0.019	—
Methane	146	4.67	2.21	2.09	0.75
	50	—	1.06	—	—
	30	0.94	0.87	—	—
	13	—	0.40	—	—
	6.1	0.21 = LMIE	—	—	—

losses from small diameter electrodes did not allow some combinations of low capacitance and high voltage at which smaller LMIEs might otherwise have been found. Table A-4-1.3b, based on data from [62,63], shows the minimum voltage across the spark gap needed to ignite the same hydrogen and methane mixtures in air. It shows that at small capacitance, at least 1000 volts is needed for hydrogen ignition and at least 3200 volts for methane ignition. Minimum ignition voltages decrease with decreased electrode diameter and with increased storage capacitance. For reference, a 12-oz beer can has a capacitance of about 6 pF, a person about 146 pF, a steel drum 30–50 pF, an automobile 500 pF and a road tanker about 1000 pF [148].

A maximum voltage of 100 V is sometimes used as a criterion for avoiding ignition by static sparks. However this criterion is very conservative. Table A-4-1.3b shows that spark ignition by an essentially noninductive, low capacitance circuit usually requires several kilovolts. A more realistic criterion for avoiding ignition of most flammable mixtures is a maximum voltage of 1000 V. It is important to note that larger capacitances allow ignition at smaller voltages. Although such large equipment is normally grounded via hard piping, a 3 m diameter tank with insulating lining may have a capacitance as high as 100,000 pF [148]. If a low voltage circuit contains a large inductance, rapid separation of contacts may result in high voltage sparks

Table A-4-1.3b : Minimum Ignition Voltage Variation with Capacitance and Electrode Diameter

	Capacitance (pF)	15 mm diameter	1.59 mm diameter	0.5 mm diameter	Points
Hydrogen	10000	—	—	—	100
	1000	—	—	—	300
	146	2300	1900	1700	1000
	50	—	2500	—	—
	30	2500	2300	—	1500
	6.1	2700	2300	2500	—
Methane	10000	—	—	—	800
	146	8000	5500	5350	3200
	50	—	6500	—	—
	30	7900	7600	—	—
	6.1	8300	—	—	—

even if the open circuit voltage is only a few volts. This is due to transient back-voltage created by the inductance, which may rise to kilovolts depending on the product of inductance and the rate of change of current as the contacts separate. Low voltage, low inductance circuits are capable of producing incendive arcs when the circuit is broken, provided a relatively large current source (exceeding "static" charging currents) is present. Insulating flanges are therefore used to prevent break-sparks in systems containing cathodic protection systems, or which are otherwise subject to stray currents.

Assume Figure A-4-1.3 represents a road tanker being loaded. The charging current I_C carried by the liquid flows onto the truck body, which has capacitance C. The truck has leakage resistance R_L to ground via its tires and loading circuit through which a leakage current I_L flows to ground. The voltage V across capacitor C varies with time t according to

$$V = I_C R_L \{1 - \exp(-t/R_L C)\} \quad (1)$$

The maximum voltage and energy occur when C is fully charged. Since all the charging current now bypasses the capacitor, the exponential term in (1) approaches zero ($t >> R_L C$) and $I_L = I_C$. Hence

Maximum Voltage $\quad V_{max} = I_C R_L \quad (2)$

Maximum Charge $\quad Q_{max} = I_C R_L C \quad (3)$

Maximum Energy $\quad W_{max} = 0.5 Q_{max} V_{max} = I_C^2 R_L^2 C/2 \quad (4)$

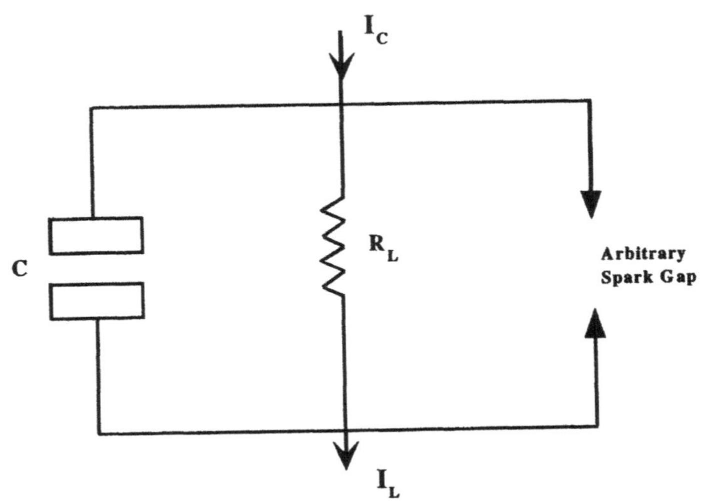

FIGURE A-4-1.3. Leaky capacitor charged with constant current I_C.

Equation (4) leads to a criterion for leakage resistance below which a specified energy W cannot be attained:

$$R_L < I_C^{-1}(2W/C)^{0.5} \tag{5}$$

To apply these data and equations to the problem of ground resistance, the maximum anticipated current must first be estimated. For practical industrial situations, I_C varies in the range 0.01–100 μA. The upper value represents extreme cases such as microfiltration and the lower value to slow flow in pipe. Typical charging currents for tank truck loading are of the order 1 μA (5-3.1.1). As an example, consider a system such as a tank with a capacitance less than 1000 pF. First, consider the minimum ignition voltages in Table A-4-1.3b. From Eq. (2), $R_L = V_{max}/I_C$. In the case of hydrogen the minimum ignition voltage at a capacitance less than 1000 pF is 300 V, so the required ground resistance must be less than 300 MΩ at 1 μA and less than 3 MΩ at 100 μA. As another example, Eq. (4) can be used to find the required ground resistance for a person having capacitance of 146 pF. Since the lowest MIE using pointed electrodes is 0.073×10^{-3} J and energy $W_{max} = I_C^2 R_L^2 C/2$, it follows that for hydrogen the ground resistance must be less than 1000 MΩ at 1 μA and less than 10 MΩ at 100 μA.

Apart from unusual circumstances discussed in 6-4.1.1, a ground resistance of less than 1 MΩ is satisfactory in practical cases even for hydrogen, which has an unusually low MIE, and less than 100 MΩ is satisfactory for moderate charging currents. For most other gases and dusts these grounding requirements for static electricity are conservative owing to their greater MIEs. Where lower MIEs may be encountered, such as the handling of anesthetic gases in oxygen enriched atmospheres, grounding requirements are more stringent as given in NFPA 99.

A-5-1.4.4. MIE of Liquid Mists

A minimum ignition energy model was derived [64] based on the "Simple Chemically Reacting System" or SCRS [65], which defines a dimensionless "mass transfer driving force" B to represent the ratio of (excess enthalpy in the bulk of the gas adjacent to the droplet surface) to the (enthalpy increase of the liquid vaporizing). The SCRS model yielded predictions of the effect of B and droplet diameter on the ignition energy of liquid mists. Experiments were performed to test the model using a series of fuel mists [64].

The minimum ignition energy expression is

$$\text{MIE (mJ)} = \{(\pi/6)c_{p,a}\Delta T_{St} D^3/\rho_a^{0.5}\} * \{\rho_f/[(\phi \ln(1+B))]\}^{3/2} \tag{1}$$

where

$c_{p,a}$ = specific heat of air at constant pressure
ΔT_{St} = temperature rise for stoichiometric combustion
D = surface average particle diameter
ρ_a = air density
ρ_f = fuel density
ϕ = equivalence ratio
B = mass transfer number

The mass transfer number B represents the ratio of the energy available for vaporization to the energy required for vaporization, and may be thought of as a driving force for mass transfer. It can be expressed as

$$B = c_{p,a}(T_{St} - T_{BP})/\Delta H \qquad (2)$$

The stoichiometric flame temperature (T_{St}) is used to characterize the burning gas surrounding the droplets because combustion naturally predominates at a distance where the fastest burning mixture is produced. This mixture approximates to the stoichiometric composition. The selection of the droplet surface temperature T_{BP} is discussed below. The enthalpy change for vaporization ΔH is given by

$$\Delta H = L + c_{p,f}(T_{St} - T_f) \qquad (3)$$

where L = heat of vaporization and T_f = initial fuel temperature

The dimensionless parameter B increases with "volatility" and typically lies between 1.2 and about 8 as shown in the following table [64]:

Liquid Hydrocarbon Fuel	Mass Transfer Number *B*
Heavy fuel oil	1.5
Light fuel oil	2.5
Diesel oil	2.8
Gas oil	3.1
Kerosene	3.7
Isooctane	6.1

A problem with the overall approach for liquid mixtures is that suitable averages must be used when calculating B, although errors in B are partly offset by the logarithmic term in Eq. (1). It is also necessary to decide at what temperature the properties of air should be evaluated. In [66] it was suggested that $c_{p,a}$ and ρ_a should be evaluated at the arithmetic mean of the ambient air and stoichiometric flame temperatures, or about 1300 K. Also, since ignition limits of fuel mixtures appeared to correlate well with the 10% evaporation point, this was taken as the average fuel boiling point when calculating *B*. Figure 5-1.4.4 shows experimentally verified ignition energies for three of the above fuels as a function of *D*, all at an equivalence ratio of 0.65.

The model assumes that liquid evaporation is always the rate controlling step. At some point the model must fail, since as droplet size approaches zero the predicted MIE approaches zero rather than the MIE of the vapor in air. In practice, droplets having diameters less than 10–40 μm completely evaporate ahead of the flame and burn as vapor (5-1.3). The model also predicts that the MIE continuously decreases as equivalence ratio is increased, although as discussed above, combustion around droplets is not restrained by the overall stoichiometry and naturally predominates at the stoichiometric concentration. It is recommended that the model be applied only to droplet diameters above about 20 μm and equivalence ratios less than about one.

A-5-2.4.1. Hyperbolic Relaxation

Equation (2-3.7) suggests that at very low values of conductivity ($\kappa \sim 0.01$ pS/m), charge will relax extremely slowly from a liquid. Filters for example would have to be an hour or more upstream of tanks before the charge would dissipate to 5% of its initial value.

Table (a) shows experimental data [24] for the initial charge density exiting a fuel filter Q_0 plus the charge density Q remaining 30 s downstream. At low conductivity the charge decays much faster than predicted by an exponential relaxation law [Eq. (2-3.7)] and instead follows a hyperbolic relaxation law [24] given by

$$Q_t = Q_0/[1 + \mu\, Q_0 t/(\varepsilon_0\, \varepsilon_r)] \qquad \text{(a)}$$

where the apparent charge carrier mobility μ (fitted to test data) is about 1×10^{-8} $m^2V^{-1}s^{-1}$ and Q_t is the charge density at time t in C/m^3. To be equivalent to a criterion of "three relaxation times" for charged liquid to relax to nonhazardous levels, the ratio $Q_t : Q_0$ in (a) should be 0.05. Rearranging (a) and substituting for μ and ε_0 gives [9]

$$t = 0.0168\, \varepsilon_r/Q_0 \text{ (s)} \qquad \text{(b)}$$

Table (b) shows charge dissipation times to reduce the charge density to 5% of the initial values for a series of initial charge densities (Q_0) and for

TABLE (a). Experimental Data for Charge Density Q Downstream of a Microfilter

k (pS/m)	Q_0 ($\mu C/m^3$) (Measured)	Q @ 30 s (Measured)	Percent remaining (Measured)	Percent predicted (Hyperbolic)	Percent predicted (Exponential)
0.0167	64.6	32.3	50	48	98
0.0205	78.2	41.4	53	43	97
0.0223	78.0	30.4	39	43	97
0.0252	75.5	28.7	38	44	96
0.0980	60.9	26.2	43	49	86
0.130	65.3	21.5	33	48	82
0.247	253	70.8	28	19	69
0.305	188	54.5	29	24	63
2.07	332	33.2	10	15	5
2.21	257	42.1	16	19	4

TABLE (b). Charge Dissipation Times for Hyperbolic Relaxation (ε_r = 2–4)

Q_0 ($\mu C/m^3$)	Time (s) to 5% of Q_0	Time (s) to 20 $\mu C/m^3$	Time (s) to 30 $\mu C/m^3$
5000	7–13	88–176	59–118
2000	17–34	87–175	58–117
1500	22–45	87–175	58–116
1000	34–67	87–174	57–114
500	67–134	85–170	55–111
100	336–672	71–142	41–82

liquid dielectric constants varying from 2 to 4. Also shown are the dissipation times needed to reduce hazardous charge densities to the "safe" threshold of <20–30 $\mu C/m^2$. Apart from the outside case of a relatively high dielectric constant and a low inlet charge density requirement, 100 s of charge dissipation usually reduces the charge density to safe values.

Owing to the change from exponential to hyperbolic relaxation, a nominal "dissipation time" of 100 s is assigned in Appendix B to all liquids whose conductivities are below 2 pS/m. The 2 pS/m demarcation is convenient since three relaxation times at $\varepsilon_r = 2$ is equivalent to 30 s or less. This makes the "30 s upstream filter placement" recommendation of API 2003 and NFPA 77 in line with the worst case BS 5958 "three relaxation times upstream" recommendation. BS 5958 recommends 100 s of relaxation time below 2 pS/m. As shown in Table (b), the time taken to reduce charge densities above 100 $\mu C/m^3$ to a safe level (<20–30 $\mu C/m^3$) via hyperbolic relaxation is roughly independent of initial charge density [8]. The same dissipation time is needed in each case. The result is significant in relation to filter placement upstream of noninerted containers. It might be argued that pipeline flow charging is much reduced at very low conductivities owing to a shortage of ions and this compensates for long charge dissipation times. However, it can be seen that to achieve safe charge density levels downstream of filters such a margin of safety does not exist and about 100 s of relaxation is still required. Charge densities can exceed 500 $\mu C/m^3$ downstream of filters even at conductivities of the order 0.01 pS/m [24].

A-5-3.5. Filtration

While attempts have been made to model the charging process in filters [122–126], the following figure [8] shows that not only are large changes of charge density possible with changes in nominal pore size, but the polarity of charge generated can change even for a single filter type (in this case a polyester bag filter) and a single liquid (Jet-A kerosene from a single tank, the

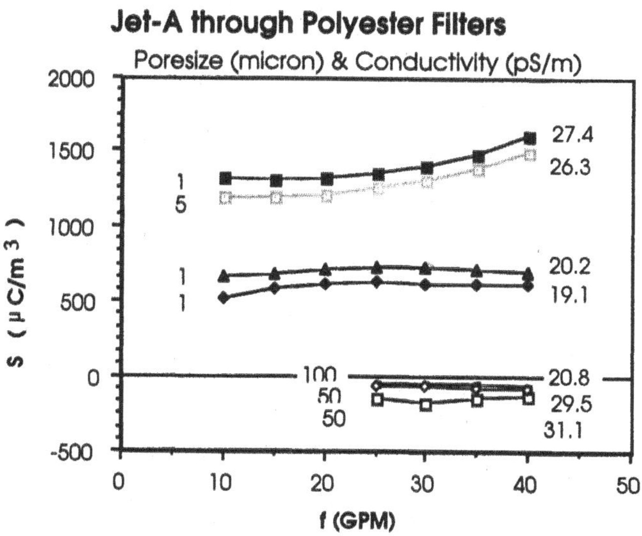

minor variations in conductivity being caused by recirculation and changes in ambient temperature). It is unlikely that polarity changes can be accounted for by any simple model. The figure shows that charge density has only a small dependence on flow rate. The charging current has an approximately proportional dependence since it is the product of charge density and flow rate.

A-5-4. Filling Criteria for Tank Operations

The following background discussion applies principally to tank filling of nonconductive liquids, whose relaxation times are sufficiently long for charge to accumulate. For any geometry and liquid fill depth of a conductive tank, the complete internal potential and electric field distributions can be calculated provided the charge density in the liquid is uniform and known. Liquid charging currents may be estimated for various inlet pipe diameters and flow rates based on empirical relationships (5-3.1.1). The rate of charge relaxation can be estimated based on liquid relaxation time. Hence, the time-dependent liquid depth and charge density can be used to solve Poisson's equation in the liquid, yielding potential and field distributions. If the vapor space is assumed to contain zero charge, the potential and field distributions can also be solved for the vapor space using the Laplace equation. Solutions are given in [69] for rectangular tanks and [8,70] for cylindrical tanks. While it is assumed that liquids approximate to uniform charge density during tank filling, the liquid surface will in reality contain excess charge since it is not in contact with a grounded surface. This and other nonidealities are typically neglected in modeling. The essential problem is to determine the minimum liquid surface potential at which ignition of a gas mixture of specified MIE can occur given the presence of an ideal electrode. Scaling models may then be used to define a maximum acceptable charging current for any size and shape of tank for any given liquid relaxation time. Empirical data may then be used to define a maximum liquid flow rate for any pipe diameter that will produce less than the maximum acceptable charging current. Various experimental and theoretical studies have been made.

In order to establish safe values for velocity–diameter product, various studies have been made to determine the minimum liquid surface potential that will result in an incendive discharge in the presence of a grounded electrode. Studies reviewed in [8] showed that for credible charging conditions, liquids must be negatively charged to yield incendive brush discharges. The consensus has been that to avoid incendive discharges the maximum liquid surface potential, which in simple tanks is produced at the geometric center of the liquid surface, should be less than (–) 25 to 60 kV. These values were based on the minimum surface potentials producing brush discharges capable of igniting mixtures of aliphatic hydrocarbons in air. Smaller scale studies led to values at the low end of this range while large scale studies led to values at the high end.

The large–scale tests involved a 1.6-m-diameter cylindrical tank containing about 2 m^3 of kerosene with conductivity 0.5 pS/m [68]. These tests suggested a minimum 58 kV surface potential for ignition, which is at the high end of the reported range. Possible reasons for the factor-of-two disparity with small scale studies are outlined in 3-8.1.2. A 15-cm-diameter cylindrical acrylic test chamber containing a grounded electrode was supported above the center of the liquid surface and 5% propane in air was supplied at a rate of 10 L/min. This concentration was considered the LMIE composition and was held constant. The electrode had a constant gap of 20 mm from the liquid surface. Although brush discharges were observed above about 25 kV, ignitions were not obtained below 58 kV. These reported surface potentials were calculated for charged liquid with no electrode present, while the ignition tests were conducted with an electrode present. Since charged liquid will discharge to an electrode as soon as breakdown conditions are reached, the significance of a higher "calculated" surface potential corresponding to ignition is unclear. As the charging current to the tank was increased, leading to an increased charge density in the tank and a correspondingly higher "calculated" surface potential, the actual surface potential in the ignition chamber should always have been limited by the gap breakdown field. As the charging current was increased, discharges should have occurred more frequently and liquid surface deformation in the electric field [71] should have decreased owing to the higher discharge frequency. This may have offset practical test limitations caused in part by the high solubility of propane in kerosene (leading to depletion of propane in the test chamber and a very lean mixture near the liquid surface), and liquid surface perturbations (bridges and cones) whose formation might affect charge transfer in the brush discharges (3-8.1.2). In one small scale study, gas solubility was overcome by starting with a rich mixture and allowing the optimum composition to appear as the test gas dissolved [71]; the charging current was adjusted so that brush discharges occurred every few seconds and would encounter the most easily ignited mixture at least once during each test cycle. The high discharge frequency prevented bridge and cone formation at the surface.

A contrary position was taken in [190]. It was argued that the ignitions observed at less than 58 kV in small scale tests were the result of excessive surface capacitance and values of about 60 kV apply to large-scale equipment. While a compelling argument was made, the fact remained that brush discharges described as "nonincendive sparks" had been observed during the large scale tests at calculated surface potentials above 25 kV [68]. The essential argument of [190] therefore became that these discharges should be regarded as harmless not only at their inception voltage of about 25 kV but also at voltages twice as large.

Since only limited ignition studies have been made, none using test gases more easily ignitable than propane or completely satisfactory with respect to test conditions, it is prudent to use the conservative value of 25 kV for general safety analyses. With small diameter electrodes (7–12 mm), brush discharges begin to be observed at approximately this surface potential. There are probabilistic factors involved not only in charge transfer for individual brush discharges but also the power density distribution within the discharge and the probability of ignition, even with all test conditions held as constant as possible. In a practical tank system containing nonidealities such as a readily mobile excess surface charge (including transient convected "spikes"), moving electrodes created during sampling and gauging, surface froth and suspended charged mist, model assumptions become less applicable and a large margin of safety is needed when applying the concept of a minimum surface potential for ignition.

A-5-4.2.1. Effect of Road Tires

Electrostatic charge caused by separation of nonconductive tire treads from the pavement can raise a tank truck potential up to 100 kV, depending on road surface roughness, vehicle speed and tire tread resistance [106]. This typically occurs on dry roads when tire treads have a resistivity of 10^9–10^{12} Ω-m. When the road surface is a good insulator, such as asphalt in dry weather, both the tank truck and the road surface can retain high potentials for significant periods after the truck has stopped. Lowering the tire resistance to below 10^9 Ω avoids the hazard of high truck potentials. "Nonstatic" tires were shown both to decrease the peak potential on the truck and increase the rate of charge dissipation once the truck came to rest.

Resistances to ground were measured for a small van with four tread 4 polyester, 2 nylon ply tires (sidewall 6 polyester plies) showing the effect of ground surface [107]. The resistance at a source potential of 50 V was measured either to an aluminum plate under one tire or to a nearby fire water line. The weather was clear and dry (73°F, 41% RH) with no rain for four days. Results are shown in the table below:

Surface	**Resistance (Ω)**
Asphalt	2×10^{10}
Concrete	5×10^{8}
Aluminum plate (one tire)	4×10^{8}
Clay/hard packed gravel	3×10^{8}
Grass	4×10^{7}

Except on dry asphalt, all the ground resistances were 5×10^8 Ω or less with the tire used. The decay time constant from an initial 300 V on the van was much less than 1 s except in the case of dry asphalt, which was several seconds. For a tank truck with typically about a dozen tires it was concluded that the resistance would be about one-third of those measured and that grounding of the truck was primarily dependent on ground surface except where independent bonding/grounding is carried out [107]. This is in agreement with an earlier review [158] which concluded that as a tank truck slows down, the equilibrium voltage decreases to relatively low values; immediately after stopping, potentials in the range 2300–5000 V have been measured. Charge drains rapidly through the tires to ground and potentials decrease to a nonhazardous level in about 10 s. There is negligible practical hazard from charging via road tires and in 1953 the requirement for drag chains was deleted from NFPA 385.

A-5-6.1. Potentials during Water Washing of Tanks

This topic received comprehensive study in the 1970s following a series of explosions during supertanker washing. A large selection of relevant literature is provided in [74–103 and 213–215]. Marine tanker and barge chemical shipments involve much smaller tanks than the center tanks involved in the supertanker explosions, and during water washing these will achieve much smaller space potentials. Marine chemical tanker operations are addressed in [104,105]. Barge tanks, owing to their small height, do not represent a credible risk during water washing provided steam is not injected directly to the tank and proper bonding and grounding is in place.

The maximum potential in a rectangular tank containing a uniform charged mist occurs at the geometrical center, and for tanks of equal volume this potential maximizes in the case of a cube. Solution of Poisson's equation (C-1.4) for a grounded, cubic tank of side "a" meters containing charged mist of uniform charge density "Q_V" C/m^3 has been carried out using finite element modeling [97] resulting in the simplified expression for tank center space potential (TCSP):

$$\text{TCSP} = 0.056 Q_V a^2/(\varepsilon_r \varepsilon_0) \qquad \text{(A-5-6.1)}$$

Direct injection of steam to a tank can greatly elevate potentials and must be avoided. Where hot water is obtained via steam injection to a mixer, the temperature should be carefully controlled particularly during start-up. For water washing, assuming a mist charge density of 3×10^{-8} C/m^3, as observed during measurements on supertankers, a cubic tank side of 10 m and dielectric constant (ε_r) of unity, the maximum TCSP via Eq. (A-5-6.1) is 19 kV. The ignition probability is related to various factors such as the maximum size of water slug (hence capacitance), the ignition energy of flammable vapor in the tank, and the fraction of the TCSP attainable by the slug. Analyses [104,105] indicate that TCSP hazard thresholds (scaled to a 3000-m^3 tank) range from about 15 kV to about 40 kV depending on the flammable liquid involved; washing hazards due to charged water mist in the absence of steam are only significant in tanks having a volume greater than 1000 m^3. Steady-state mist charge densities tend to be larger in smaller tanks and can attain values in the range 100–250 nC/m^3. However, both the relatively small volume and noncubic geometry makes water washing hazards negligible in the case of barge tanks. Most analyses such as [104,105,213] include considerations of water slug size and maximum spark energy, from which it can be concluded that the hazard is negligible in all tanks less than 100 m^3. Documented incidents not involving extraneous factors such as insertion of ungrounded hoses have all occurred in tanks having volumes greater than 1000 m^3.

A far more conservative approach for industrial tanks was taken in [127] in which the requirement for a water slug was omitted from the analysis. Instead it was assumed that if the electric field at some projection in the charged water mist exceeds the breakdown field of air, a brush discharge capable of igniting flammable gas might be produced. The results of the analysis are summarized in 5-6.1.1 and involve a tank size limit of 30 m^3 or diameter of 3 m. It should be noted that this scenario was considered very unlikely in [86] and other experimental studies related to tanker washing [74-103]; for example, wet surfaces tend to give corona rather than brush discharges.

A-6-1.2. Effect of Particle Size on Dust MIE

As in the case of mists (A-5-1.4.4), the MIE of dusts varies roughly with the cube of the surface average diameter (*D*) of the suspension [128,129], although in some cases the power dependence for dusts is found to be second order or fractional. Solid particle combustion may be considered in terms of two groups. The first of these (including Ti, Zr, B, Si and C) burns in the solid phase since the element has a higher boiling point than its oxide. In the second (Mg, Al) the metal vaporizes at a lower temperature than its oxide and combustion occurs in the gas phase. Dusts may burn in both phases where more than one oxide may form or where there is limited evolution of volatiles (some coal dusts). The great importance of particle diameter (*D*) is shown in Figure A-6-1.2 for atomized aluminum, adapted from [128].

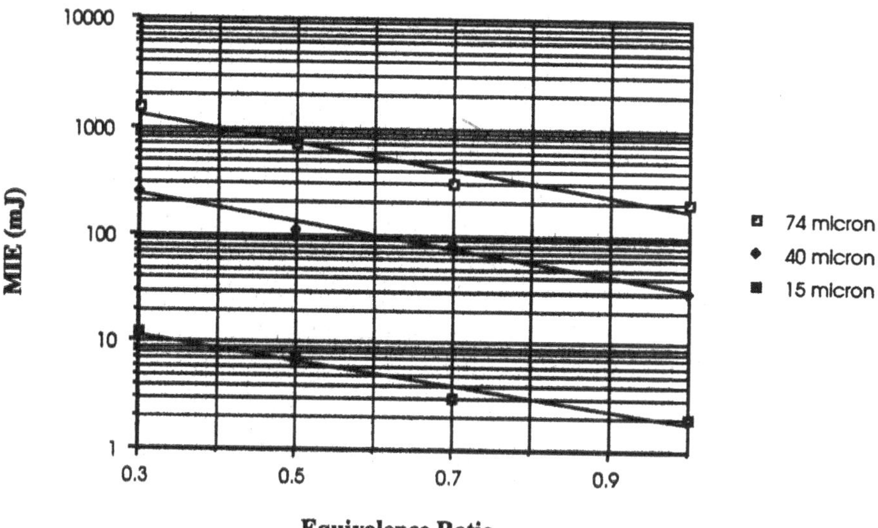

FIGURE A-6-1.2. Schematic effect of particle size and equivalence ratio on MIE of aluminum.

A-6-1.3.1. Ignition Energy of Hybrid Mixtures

The general function giving linear plots on semilogarithmic graphs is:

$$y = df^{gx} \tag{1}$$

where d and f are positive constants and g is any constant.

Rewriting

$$y = d(f^g)^x \tag{2}$$

$$\ln(y) = \ln(d) + (\ln f^g)x \tag{3}$$

Let

$$Y = \ln(y), \quad a = \ln(d), \quad b = \ln(f^g)$$

so

$$Y = a + bx \tag{4}$$

To apply these relationships to the hybrid mixture MIE problem, it is noted that only two points need be defined on the Y axis: the MIE of the dust in air and the LMIE of the gas in air. The first on the x axis corresponds to zero gas, so $x_1 = 0$, and the second to the optimum gas concentration. All the unknowns are experimental quantities.

Since

$$Y = \ln y = a + bx$$

$$\text{For dust:} \quad \ln(y_1) = a + b(x_1)$$

$$\text{For gas:} \quad \ln(y_2) = a + b(x_2)$$

Since $x_1 = 0$,

$$a = \ln(y_1) \tag{5}$$

$$b = -\ln(y_1/y_2)/x_2 \tag{6}$$

Hence the intercept a and slope b have been determined for the equation $Y = a + bx$. Since $Y = \ln(y)$:

$$\ln(y) = \ln(y_1) - (x/x_2)\ln(y_1/y_2)$$

or

$$y = \exp[\ln(y_1) - (x/x_2)\ln(y_1/y_2)] \tag{7}$$

The right-hand side of this equation contains only test data plus the gas concentration of interest (x). Test data are the dust MIE in air (y_1), the gas MIE in air (y_2) and the optimum gas concentration (x_2).

A-6-1.5. Effect of Temperature on Powder MIE

The general function giving straight lines on double logarithmic plots is:

$$y = cx^b \tag{1}$$

or

$$\ln y = \ln c + b \ln x$$

Let

$$Y = \ln y, \qquad a = \ln c, \qquad X = \ln x$$

Hence

$$Y = a + bX \tag{2}$$

Solving for the constants a and b and substituting logarithmic terms one obtains:

$$y = \exp\{\ln y_1 + \ln(x/x_1) * \ln(y_1/y_2)/\ln(x_1/x_2)\} \tag{3}$$

This equation can be used to estimate the effect of temperature on MIE as discussed in 6-1.5:

$$\text{MIE at temperature } T \text{ (Celsius)} = \exp\{\ln M + \ln(T/20) * \ln(M/0.088)/\ln(20/1000)\} \tag{4}$$

where M = MIE measured at 20°C

APPENDIX

B

DATA TABLES

These tables are adapted from Reference [9]. See footnotes for explanations of symbols.

Flammability Data for Gases and Vapors in Dry Air at 1 atm, 25°C Except Where Indicated

Gas/Vapor	LMIE = G (mJ)	Stoi (vol%)	Opt = C_0 (vol%)	LFL (vol%)	UFL (vol%)	Su (cm/s)	LOC in N_2 (vol%)
acetaldehyde	0.37	7.73		4.0	57.0		
acetone	1.15	4.97	4.5	2.6	12.8	54	11.5
acetone (in oxygen)	0.0024			2.5	60		n/a
acetylene	0.017	7.72	8.5	2.5	100	166	0
acetylene (in oxygen)	0.0002	28.6	40	2.5	100	1140	n/a
acetylene (decomp @ 1 atm)	10^3–10^5	100	100	n/a	100		0
acetylene (decomp @ 2 atm)	100–1000	100	100	n/a	100		0
acrolein	0.13	5.64		2.8	31	66	
acrylonitrile	0.16	5.29	9.0	3.0	17.0	50	
allyl chloride	0.77			2.9	11.1		
ammonia	680	21.8		15	28		
benzene	0.20	2.72	4.7	1.2	7.8	48	11.4
1,3 butadiene	0.13	3.67	5.2	2.0	12	68	10.5
n-butane	0.25	3.12	4.7	1.6	8.4	45	12
n-butane (in oxygen)	0.009	13.3		1.6	49		n/a

Gas/Vapor	LMIE = G (mJ)	Stoi (vol%)	Opt = C_0 (vol%)	LFL (vol%)	UFL (vol%)	Su (cm/s)	LOC in N_2 (vol%)
n-butyl chloride	1.24	3.37		1.8	10.1		
carbon disulfide	0.009	6.53	7.8	1.0	50.0	58	5
cyclohexane	0.22	2.27	3.8	1.3	8.0	46	
cyclopentadiene	0.67					46	
cyclopentane	0.54	2.71		1.5		44	
cyclopropane	0.17	4.44	6.3	2.4	10.4	56	11.5
cyclopropane (in oxygen)	0.001		17	2.5	60		n/a
dichlorosilane	0.015	17.4		4.7	96		
diethyl ether	0.19	3.37	5.1	1.9	36	47	10.5
diethyl ether (in oxygen)	0.0012		14	2.0	82		n/a
dihydropyran	0.36						
diisobutylene	0.96			1.1	6.0		
diisopropyl ether	1.14			1.4	7.9		
dimethoxymethane (methylal)	0.42			2.2	13.8		
2,2-dimethylbutane	0.25	2.16	3.4	1.2	7.0	42	
dimethyl ether	0.29			3.4	27.0	54	
2,2-dimethyl propane	1.57			1.4	7.5	39	
dimethyl sulfide	0.48			2.2	19.7		
di-*t*-butyl peroxide	0.41						
ethane	0.24	5.64	6.5	3.0	12.5	47	11
ethane (in oxygen)	0.0019	22.2	17	3.0	66		n/a
ethanol		6.53		3.3	19		10.5
ethyl acetate	0.46	4.02	5.2	2.0	11.5	38	
ethyl acrylate				1.4	14		8
ethylamine	2.4	5.28		3.5	14		
ethylene	0.084	6.53	7.5	2.7	36	80	10
ethylene (in oxygen)	0.00094	25.0	22.5	3.0	80		n/a
ethyleneimine	0.48			3.6	46	46	
ethylene oxide	0.065	7.72	10.8	3.0	100	108	0
ethylene oxide (decomp)	~1500	100	n/a	n/a	100		0

Gas/Vapor	LMIE = G (mJ)	Stoi (vol%)	Opt = C_0 (vol%)	LFL (vol%)	UFL (vol%)	Su (cm/s)	LOC in N_2 (vol%)
furan	0.22	4.44		2.3	14.3		
n-heptane	0.24	1.87	3.4	1.05	6.7	46	11.5
n-hexane	0.24	2.16	3.8	1.1	7.5	46	12
n-hexane (in oxygen)	0.006	9.52		1.2	52*		n/a
hydrogen	0.016	29.5	28	5±1	75	312	5
hydrogen (in oxygen)	0.0012	66.7		4.0	94	1400	n/a
hydrogen (in nitric oxide)	8.7	50.0					n/a
hydrogen sulfide	0.068			4.0	44		7.5
isooctane	1.35	1.65		0.95	6.0	41	
isopropyl alcohol	0.65	4.44		2.0	12.7	41	
isopropyl chloride	1.08			2.8	10.7		
isopropylamine	2.0					31	
methane	0.21	9.47	8.5	5.0	15.0	40	12
methane (in oxygen)	0.0027	33.3	23	5.1	61	450	n/a
methane (in nitric oxide)	8.7						n/a
methanol	0.14	12.24	14.7	6.0	36.0	56	10
methylacetylene	0.11	4.98	6.5	1.7			
methyl acrylate				2.8	25		8.7
methylene chloride	>1000			14	22		19
methylene chloride (in oxygen)	0.137			11.7	68		n/a
methyl ethyl ketone	0.53	3.66	5.3	2.0	12.0		11
methyl butane (isopentane)	0.21	2.55	3.8	1.4	7.6	43	12
methyl cyclohexane	0.27	1.96	3.5	1.2	6.7	44	
methyl formate	0.4			4.5	23		
n-pentane	0.28	2.55	3.3	1.5	7.8	46	12
2-methyl pentene	0.18	2.72	4.4			47	
propane	0.25	4.02	5.2	2.1	9.5	46	11.5
propane (in oxygen)	0.0021	16.7	15				n/a
propionaldehyde	0.32			2.6	17	58	

Gas/Vapor	LMIE = G (mJ)	Stoi (vol%)	Opt = C_0 (vol%)	LFL (vol%)	UFL (vol%)	Su (cm/s)	LOC in N_2 (vol%)
n-propyl chloride	1.08			2.6	11.1		
propylene	0.28	4.45		2.0	11.0	52	11.5
propylene oxide	0.13	4.98	7.5	2.3	36.0	82	7.8
isopropyl mercaptan	0.53						
styrene		2.05		0.9	6.8		9.0
tetrahydrofuran	0.54			2.0	11.8		
tetrahydropyran	0.22		4.7			48	
thiophene	0.39						
toluene	0.24	2.27	4.1	1.1	7.1	41	9.5
trichloroethane				7.5	12.5		14
trichloroethane (in oxygen)	0.092			5.5*	57*		n/a
trichlorosilane	0.017			7.0	83		
triethylamine	0.75	2.10		1.2	8.0		
vinyl acetate	0.7	4.45		2.6	13.4		
vinyl acetylene	0.082			1.7	100		0
xylene(s)	0.2	1.96		1.0	7.0		

LMIE = lowest minimum ignition energy measured at optimum concentration [*G* in Eq. (6-1.3.1)]
Stoi = stoichiometric concentration
Opt = optimum concentration at which LMIE was measured [C_0 in Eq. (6-1.3.1)]
LFL = lower flammable limit
UFL = upper flammable limit
Su = fundamental burning velocity
LOC = limiting oxygen concentration of fuel–oxygen–nitrogen mixture
n/a = not applicable

Note 1: It is possible that the fuel concentration most easily ignited by capacitive spark discharge, as reported in this table (C_0), might differ from that most easily ignited by other types of static discharge (eg brush discharge). See 2-6.2.1.

Note 2: LFL and UFL values in air taken primarily from NFPA 325.

Note 3: LMIE, LFL and UFL values in oxygen taken primarily from NFPA 53 (* denotes elevated temperature).

Note 4: Su values taken from [223].

Note 5: LOC values taken primarily from NFPA 69.

Typical Conductivities, Dielectric Constants and Relaxation (or Dissipation) Times of Liquids

Conductive Liquids ($\kappa > 10^4$ pS/m)

Liquid	κ (pS/m)	ε_r	$\tau = \varepsilon_0\varepsilon_r/\kappa$ (s)
acetaldehyde (15°C)	1.7×10^8	21.1	1.1×10^{-6}
acetamide	8.8×10^7	59	5.9×10^{-6}
acetic acid (0°C)	5×10^5	6.15	1.1×10^{-4}
acetic acid (25°C)	1.12×10^6	6.15	4.9×10^{-5}
acetic anhydride (25°C)	4.8×10^7		
acetone (25°C)	6×10^6	20.7	3×10^{-5}
acetonitrile (20°C)	7×10^8	37.5	5×10-7
acetophenone (25°C)	3.1×10^5	17.39	5.0×10^{-4}
acetyl bromide (25°C)	2.4×10^8		
acetyl chloride (25°C)	4×10^7		
acrolein	1.55×10^7		
acrylonitrile	7×10^5	38	4.8×10^{-4}
allyl alcohol (25°C)	7×10^8		
aminoethylethanolamine	$> 1 \times 10^6$		
n-aminoethyl piperazine	2.4×10^5		
ammonia (–79°C)	1.3×10^7		
iso-amyl alcohol	1.4×10^5	14.7	9.3×10^{-4}
aniline (25°C)	2.4×10^6	6.89	2.5×10^{-5}
anthracene (230°C)	3×10^4		
arsenic tribromide (25°C)	1.5×10^8		
arsenic trichloride (25°C)	1.2×10^8	n/a	n/a
benzaldehyde (25°C)	1.5×10^7		
benzoic acid (125°C)	3×10^5		
benzonitrile (25°C)	5×10^6	25.2	$4{,}5 \times 10^{-5}$
benzyl alcohol (25°C)	1.8×10^8		
benzylamine (25°C)	$<1.7 \times 10^6$		
benzyl benzoate (25°C)	$<1 \times 10^5$		
benzyl cyanide	$<5 \times 10^6$	18.7	$> 3.3 \times 10^{-5}$
biphenyl (liquid: above 120°C)	$>1 \times 10^4$		
bromoform (tribromomethane) (25°C)	$<2 \times 10^6$	4.39	$> 1.9 \times 10^{-5}$
n-butyl alcohol (23°C)*	2.0×10^7		

Liquid	κ (pS/m)	ε_r	$\tau = \varepsilon_0\varepsilon_r/\kappa$ (s)
iso-butyl alcohol	9.12×10^5	17.51	1.7×10^{-4}
sec-butyl alcohol	$<1 \times 10^7$	16.56	$>1.5 \times 10^{-5}$
tert-butyl alcohol	2.66×10^6	12.47	4.2×10^{-5}
butyl CARBITOL	$>1 \times 10^6$		
iso-butyl chloride	1×10^4	6.49	5.7×10^{-3}
sec-butyl chloride	1×10^4	7.09	6.3×10^{-3}
capronitrile (25°C)	3.7×10^8		
m-chloroaniline (25°C)	5×10^6		
chlorohydrin (25°C)	5×10^7		
m-cresol	1.397×10^6	11.8	7.5×10^{-5}
o-cresol	1.27×10^5	11.5	8.0×10^{-4}
p-cresol	1.378×10^6	9.91	6.4×10^{-5}
cyanogen	$<7 \times 10^5$		
cyclohexanone	5×10^5		
cymene (25°C)	$<2 \times 10^6$		
dibutyl-*o*-phthalate	1.8×10^5	6.436	3.2×10^{-4}
dichloroacetic acid (25°C)	7×10^6		
cis-dichloroethylene	8.5×10^5	9.20	9.6×10^{-5}
dichlorohydrin (25°C)	1.2×10^9		
diethanolamine (20°C)*	4.6×10^7		
diethylamine (–33.5°C)	2.2×10^5		
diethyl carbonate (25°C)	1.7×10^6	2.82	1.5×10^{-5}
diethylene glycol	5.86×10^7	31.69	4.8×10^{-6}
diethylenetriamine	$>1 \times 10^6$		
diethyl ketone*	5.2×10^7		
diethyl oxalate (25°C)	7.6×10^7		
diethyl sulfate (25°C)	2.6×10^7		
dimethyl acetamide	1.1×10^7		
dimethyl formamide	6×10^6	36.71	5.4×10^{-5}
dimethyl sulfate (0°C)	1.6×10^7		
dimethyl sulfoxide	2×10^5	46.68	2.1×10^{-3}
diphenyl oxide	$<1.7 \times 10^6$	4.22	$>2.2 \times 10^{-5}$
epichlorohydrin (25°C)	3.4×10^6	22.6	5.9×10^{-5}
ethanolamine	1.1×10^9	37.72	3.0×10^{-7}
ethyl acetate (4°C)*	2.9×10^4	6.02	1.84×10^{-3}

Liquid	κ (pS/m)	ε_r	$\tau = \varepsilon_0 \varepsilon_r / \kappa$ (s)
ethyl acetate (23°C)*	4.6×10^4	6.02	1.16×10^{-3}
ethyl acetoacetate (25°C)	4×10^6	15.7	3.5×10^{-5}
ethyl acrylate	3.35×10^5		
ethyl alcohol (25°C)	1.35×10^5	24.55	1.6×10^{-3}
ethylamine (0°C)	4×10^7		
ethyl benzoate (25°C)	$<1 \times 10^5$	6.02	$>5.3 \times 10^{-4}$
ethyl bromide (25°C)	$<2 \times 10^6$	9.39	$>4.2 \times 10^{-5}$
ethyl chloride	$<3 \times 10^5$	9.45	$>2.8 \times 10^{-4}$
ethyl cyanoacetate	6.9×10^7	26.7	3.4×10^{-6}
ethylene carbonate	$<1 \times 10^7$	89.6	$>7.9 \times 10^{-5}$
ethylenediamine	9×10^6	12.9	1.3×10^{-5}
ethylene glycol	1.16×10^8	37.7	2.9×10^{-6}
ethylene glycol monobutyl ether	4.32×10^7	9.30	1.9×10^{-6}
ethylene glycol monoethyl ether	9.3×10^6	29.6	2.8×10^{-5}
ethylene glycol monomethyl ether	1.09×10^8	16.93	1.4×10^{-6}
ethyleneimine	8×10^8	18.3	2.0×10^{-7}
ethylene oxide	4×10^6	12.7	2.8×10^{-5}
ethyl formate	1.45×10^5	7.16	4.4×10^{-4}
ethylidene chloride	2.0×10^5	10.0	4.4×10^{-4}
ethyl isothiocyanate (25°C)	1.26×10^7		
ethyl lactate	1.0×108	13.1	1.2×10^{-6}
ethyl nitrate (25°C)	5.3×10^7		
ethyl oxalate	7.12×10^7	1.8 (?)	2.2×10^{-7}
ethyl propionate	8.33×10^{10}	5.65	6×10^{-10}
ethyl thiocyanate (25°C)	1.2×10^8		
eugenol (25°C)	$<1.7 \times 10^6$		
formamide (25°C)	4×10^8	111.0	2×10^{-6}
formic acid (25°C)	6.4×10^9	58.5	8.1×10^{-8}
furfural (25°C)	1.5×10^8		
glycerol (25°C)	6.4×10^6	42.5	5.9×10^{-5}
guaiacol (25°C)	2.8×10^7		
hydrogen bromide (-80°C)	8×105		
hydrogen chloride (-96°C)	1×10^6		
hydrogen cyanide (0°C)	3.3×10^8		
hydrogen iodide (@ B.P.)	2×10^7		

Liquid	κ (pS/m)	ε_r	$\tau = \varepsilon_0\varepsilon_r/\kappa$ (s)
iodine (110°C)	1.3×10^4		
mercury (0°C)	1.063×10^{18}		
methoxy triglycol	$>1 \times 10^6$		
methyl acetamide	2×10^7	191.3	8.5×10^{-5}
methyl acetate (25°C)	3.4×10^8	6.68	1.7×10^{-7}
methyl alcohol (18°C)	4.4×10^7	32.70	6.6×10^{-6}
methyl cyanoacetate	4.49×10^7	29.30	5.8×10^{-6}
methyl ethyl ketone (25°C)	1×10^7	18.51	1.6×10^{-5}
methyl formamide	8×10^7	182.4	2.0×10^{-5}
methyl formate	1.92×10^8	8.5	3.9×10^{-7}
methyl iodide (25°C)	$<2 \times 10^6$		
methyl isobutyl ketone	$<5.2 \times 10^6$	13.11	$>2.2 \times 10^{-5}$
methyl nitrate (25°C)	4.5×10^8		n/a
n-methyl-2-pyrrolidone	2×10^6	32.0	1.4×10^{-4}
methyl thiocyanate (25°C)	1.5×10^8		
monoethanolamine (20°C)*	1.2×10^9		
naphthalene (82°C)	4×10^4		
NIAX Catalyst C–271*	2.6×10^9		
nitrobenzene (0°C)	5×10^5	34.82	6.2×10^{-4}
nitroethane	5×10^7	28.06	5.0×10^{-6}
nitromethane (18°C)	6×10^7	35.87	5.3×10^{-6}
1-nitropropane	3.3×10^7	23.24	6.2×10^{-6}
2-nitropropane	5×10^7	25.52	4.5×10^{-6}
o- or *m*- nitrotoluene (25°C)	$<2 \times 10^7$		
octyl alcohol	1.39×10^7	10.34	6.9×10^{-6}
2,4-pentanedione	8.82×10^8		
phenetole (25°C)	$<1.7 \times 10^6$		
phenol	1×10^6	9.78	8.7×10^{-5}
phenyl isothiocyanate (25°C)	1.4×10^8		
phosgene (25°C)	7×10^5		
phosphorus (25°C)	4×10^8		
phosphorus oxychloride (25°C)	2.2×10^8		
pinene (23°C)	$<2 \times 10^4$		
piperidine (25°C)	$<2 \times 10^7$		
polyalkylene oxide (Y–6132)	$>1 \times 10^7$		

Liquid	κ (pS/m)	ε_r	$\tau = \varepsilon_0\varepsilon_r/\kappa$ (s)
polyalkylene oxide (Y–6854)	5.4×10^5		
propionaldehyde (25°C)	8.5×10^7	18.5	1.9×10^{-6}
propionitrile	8.51×10^6	27.2	2.8×10^{-5}
iso-propyl acetate (4°C)*	1.3×10^4		
iso-propyl acetate (23°C)*	6.5×10^4		
n-propyl acetate (23°C)*	1.12×10^4		
n-propyl alcohol (25°C)	2×10^6	20.33	9×10^{-5}
iso-propyl alcohol (25°C)	3.5×10^8	19.92	5×10^{-7}
propyl formate	5.5×10^9	7.72	1.2×10^{-8}
propyl PROPASOL	$>1 \times 10^6$		
pyridine (18°C)	5.3×10^6	12.4	2.1×10^{-5}
quinoline (25°C)	2.2×10^6	9.00	3.6×10^{-5}
salicyaldehyde (25°C)	1.6×10^7	13.9	7.5×10^{-6}
succinonitrile	5.64×10^{10}	56.5	$8.9 \times 10\text{-}9$
sulfolane (tetramethylene sulfone)	$<2 \times 10^6$	43.3	$>1.9 \times 10^{-4}$
sulfonyl chloride (25°C)	2×10^8		
sulfuric acid (25°C)	1×10^{12}		
tetraethylenepentamine	$>1 \times 10^6$		
tetrahydrofuran	3.3×10^5		
tetramethylurea	$<6 \times 10^6$	23.06	$>3.4 \times 10^{-5}$
m-toluidine	5.5×10^4	9.91	1.6×10^{-3}
o-toluidine	3.79×10^7	6.34	1.5×10^{-6}
p-toluidine (100°C)	6.2×10^6	4.98	7.1×10^{-6}
trichloroacetic acid (25°C)	3×10^5		
1,1,1-trichloroethane	7.3×10^5	7.53	9.1×10^{-5}
triethanolamine*	2.2×10^7		
triethylene glycol	8.4×10^6	23.69	2.5×10^{-5}
triethylenetetramine	$>1 \times 10^6$		
trimethylamine (–33.5°C)	2.2×10^4		
vinyl acetate (4°C)*	2.0×10^4		
vinyl acetate (23°C)*	2.6×104		
vinyltrichlorosilane	5×10^4		
water (extremely pure)	4.3×10^6	80.4	1.7×10^{-4}
water (air distilled)	$\sim 1 \times 10^9$	80.4	7.1×10^{-7}

Semiconductive Liquids ($1000 < \kappa < 10000$ pS/m)

Liquid	κ (pS/m)	ε_r	$\tau = \varepsilon_0\varepsilon_r/\kappa$ (s)
primary amyl acetate (4°C)*	1250	4.75	0.034
primary amyl acetate (23°C)*	2160	4.75	0.020
biphenyl (liquid @ 69-120°C)	2500–10,000		
bromobenzene	1200	5.40	0.040
1-bromonaphthalene	3660	4.83	0.011
iso-butyl acetate (4°C)*	2650	5.3	0.018
iso-butyl acetate (23°C)*	4320	5.3	0.011
n-butyl acetate (4°C)*	2170	5.0	0.020
n-butyl acetate (23°C)*	4700	5.0	0.0094
butyl acrylate	3580		
n-butyl propionate (24°C)*	2300		
chlorobenzene	7000	5.621	0.0071
chloroform	<10,000	4.806	>0.0043
dibutyl sebacate	1700	4.54	0.024
o-dichlorobenzene	3000	9.93	0.0029
ethylene dibromide	<10,000	4.78	>0.0042
ethylene dichloride	4000	10.36	0.022
2-ethyl hexanol*	7900		
methylene chloride	4300	8.93	0.018
pentyl acetate (23°C)*	3400		
propionic acid (25°C)*	8500	3.44	0.0036
n-propyl acetate (4°C)*	8460	8.1	0.0085
sulfur (130°C)	5000		
vinyltrimethoxysilane (<2% CH_3OH)	5900		

Semiconductive Liquids ($100 < \kappa < 1000$ pS/m)

Liquid	κ (pS/m)	ε_r	$\tau = \varepsilon_0\varepsilon_r/\kappa$ (s)
armeen	470		
2-ethylhexyl acrylate	610		
gasoline (leaded)	~100	2.3	~0.41
hydrogen sulfide (@ B.P.)	1000		
isopropyl ether	500		
sulfur (115°C)	100		
1,2,4-trichlorobenzene	200	4.08	0.18
trichloroethylene	800	3.42	0.037

Nonconductive Liquids (κ < 100 pS/m)

Liquid	κ (pS/m)	ε_r	$\tau = \varepsilon_0\varepsilon_r/\kappa$ (s)
anisole (methyl phenyl ether)	10	4.33	3.8
benzene (purified)	5×10^{-3}	2.3	~100 (dissipation)
biphenyl (solid: less than 69°C)	0.17		n/a
bromine (17.2°C)	13		
butyl stearate	21	3.111	1.3
caprylic acid (octanoic acid)	<37	2.45	> 0.58
carbon disulfide (1°C)	7.8×10^{-4}	2.6	~100 (dissipation)
carbon tetrachloride	4×10^{-4}	2.238	~100 (dissipation)
chlorine (–70°C)	<0.01		
cyclohexane	<2	2.0	>8.8
decalin	6	2.18	3.2
dichlorosilane			
diesel oil (purified)	~0.1	~ 2	~100 (dissipation)
diethyl ether	30	4.6	1.4
1,4-dioxane (diethylene oxide)	0.1	2.2	~100 (dissipation)
ethyl benzene	30	2.3	0.68
gasoline (straight run)	~0.1	~2	~100 (dissipation)
gasoline (unleaded)	<50 (varies)		
heptane (purified)	3×10^{-2}	2.0	~100 (dissipation)
hexane (purified)	1×10^{-5}	1.90	~100 (dissipation)
hexamethyldisilazane	29		
jet fuel A and A-1	0.01–50	2.2	0.39 - 100
jet fuel B	0.01–50	2.2	0.39 - 100
kerosene	1–50	2.2	0.39 - 19
methyl methacrylate			
methyl tertiary butyl ether (MTBE)			
naphthas	various	~2	~100 (dissipation)
pentachlorodiphenyl	0.8	5.06	~100 (dissipation)

Liquid	κ (pS/m)	ε_r	$\tau = \varepsilon_0\varepsilon_r/\kappa$ (s)
pentachloroethane	<100	3.83	>0.3
SiH fluid (Y-10354)	2.5		
silicon tetrachloride			
stearic acid (80°C)	<40		
styrene monomer	10	2.43	2.2
toluene	<1	2.38	~100 (dissipation)
trichlorosilane			
turpentine	22		
iso-valeric acid	<40	2.64	>0.58
xylene	0.1	2.38	~100 (dissipation)

Note 1: Conductivity varies with temperature and purity.
Note 2: Where "<" appears, test equipment probably had insufficient sensitivity to determine the proper value (e.g., cymene).
Note 3: Viscous nonconductive liquids can have exceptionally long charge dissipation times (5-2.5.4).
Note 4: Data recently measured by Union Carbide's Physical Properties Group are marked with an asterisk.
Note 5: The term "Dissipation" is explained in 5–2.5.3 and in "Glossary" under "Effective Conductivity"

Typical Resistivities, Dielectric Constants and Breakdown Strengths of Solid Dielectrics [182–183]

Solid	Resistivity (Ω-m)	Dielectric Constant	Breakdown Strength (V/mil)	Breakdown Strength (V/μm)
cellulose acetate	10^8–10^{11}	3.5–7	250–365	10–14
cellulose acetate butyrate	10^8–10^{10}	3.5–6.4		
epoxy cast resin, unfilled	10^{10}–10^{12}	3.5–5.0	320	13
glass (borosilicate)		3.8–6.8	<500	<20
melamine formaldehyde	10^{10}–10^{12}	6–9	300–400	12–16
mica (ruby)		7	1000	39
phenol formaldehyde cast resin, unfilled	10^{10}–10^{11}	4–7	250–400	10–16
polyamide (nylon)		3.6–4.6	300–470	12–18
polycarbonate	10^{14}	3.2		

Solid	Resistivity (Ω-m)	Dielectric Constant	Breakdown Strength (V/mil)	Breakdown Strength (V/μm)
polyester, glass reinforced	10^9	4.0–5.5	250–450	10–18
polyethylene	10^{17}–10^{18}	2.3	1000	39
polyethylene terephthalate (film)	10^{11}	3.1	4000–5000	160–200
polymethylmethacrylate	$>10^{12}$	3.5–4.5	380–390	15
polystyrene	10^{15}–10^{17}	2.45–2.65	500–700	20–28
polystyrene film		2.3	1000	39
polychlorotrifluoroethylene (PCTFE)	10^{16}	2.8	2500	100
polychlorotrifluoroethylene film	10^{16}	2.3–2.8	3000–5000	120–200
polypropylene	10^{15}	2.3		
polytetrafluoroethylene (PTFE)	10^{16}	2.0	480	19
polytetrafluoroethylene shaved film	10^{16}	2.0	1000–2000	39–79
polyvinylchloride (PVC)	10^9–10^{14}	6–7	400	16
quartz		3.8–4.1	1000	39
urea formaldehyde		6–7	300–400	12–16

Note 1: Resistivity varies with temperature and exact form of dielectric, including any filler used.

Note 2: Dielectric constant is typically measured using a high frequency source and depends on frequency.

Note 3: Breakdown strength is highly dependent on test method and sample preparation.

Nominal Opening Sizes of Standard Wire Cloth Sieves Commonly Used to Characterize Powders; See ASTM E11 for Complete List of US Standard Sieve Sizes

Sieve Mesh Number	Nominal Opening (μm)	Sieve Mesh Number	Nominal Opening (μm)
10	2000	230	63
40	425	270	53
120	125	325	45
200	75	400	38

APPENDIX

C

FORMULAS AND MATHEMATICAL RELATIONSHIPS (SI UNITS)

C-1. Principal Relationships

C-1.1. Capacitance (C)

Of conductors:

1. Sphere:	$C = 4\pi a\varepsilon_r\varepsilon_0$ where a = sphere radius
2. Sphere and plate:	$C = 4\pi a\varepsilon_r\varepsilon_0[0.577 + 0.5\ln(2a/d)]$ where a = sphere radius, d = sphere-plate gap
3. Coaxial cylinders:	$C = 2\pi l\varepsilon_r\varepsilon_0 / \ln(r_2/r_1)$ where l = length, r_1,r_2 = inner and outer radii
4. Parallel plates:	$C = A\varepsilon_r\varepsilon_0/d$ where A = plate area, d = gap
5. General:	$C = Q/V$
6. Series addition:	$1/C_{tot} = 1/C_1 + 1/C_2 + 1/C_3 + \ldots + 1/C_n$
7. Parallel addition:	$C_{tot} = C_1 + C_2 + C_3 + \ldots + C_n$

C-1.2. Energy (W)

Stored in capacitor:

$$W = 0.5CV^2 = 0.5QV = 0.5Q^2/C$$

C-1.3. Field (E)

See C-2.5.4 and C-2.5.5. For a charged nonconductor:

1. Field above infinite plane of uniform charge density, isolated from all other objects: $E = Q_S/2\varepsilon_r\varepsilon_0$
2. Field between infinite plane of uniform charge density and parallel plane surface: $E = Q_S/\varepsilon_r\varepsilon_0$

C-1.4. Poisson's Equation

$\nabla^2\phi = -Q_v/(\varepsilon_r\varepsilon_0)$. Analytical solutions exist for electric field and potential within simple geometrical volumes such as long grounded cylinders or grounded spheres containing material of uniform charge density (Q_V):

1. Voltage in grounded cylinder: $V = Q_V R^2[1 - (r^2/R^2)]/4\varepsilon_r\varepsilon_0$ (R = cylinder radius, r = distance from axis)
2. Maximum field (at wall) in cylinder: $E_{max} = Q_v R/2\varepsilon_r\varepsilon_0$
3. Maximum voltage (on axis) in cylinder: $V_{max} = Q_v R^2/4\varepsilon_r\varepsilon_0$, e.g., charged liquids or powders in a pipeline
4. Voltage in grounded sphere: $V = Q_v(R^2 - r^2)/6\varepsilon_r\varepsilon_0$ (R = sphere radius, r = distance from center)
5. Maximum field (at wall) in sphere: $E_{max} = Q_v R/3\varepsilon_r\varepsilon_0$
6. Maximum voltage (at center) in sphere: $V_{max} = Q_v R^2/6\varepsilon_r\varepsilon_0$ e.g., charged powder, liquid or mist in a spherical container (fair approximation for cubic containers)

C-1.5. Potential (ϕ)

Due to point charge q at distance r: $\phi = q/(4\pi\varepsilon_0 r)$

C-1.6. Resistance (R)

Proportionality of current density J to electric field E is given by Ohm's law:

1. General: $J = \kappa E$, where κ = conductivity (S/m)
2. Narrow conductors such as wires: $I = V/R$

C-2. Analysis: Grounded Sphere above Charged Nonconductive Disc

This analysis was made during preparation of a paper written jointly with Dr. T.J. Williams [71]. It illustrates a number of electrostatic principles which can be applied to the use of electric field meters and the problem of brush discharge onset. It is assumed that dielectric constant $\varepsilon_r = 1$.

C-2.1. Use of Method of Images to Find Induced Charge on Grounded Sphere

Simplification of the problem is achieved by the method of images. Consider the point charge q on the uniformly charged nonconductive surface in Figure C-2.1. This charge must induce an opposite charge on the surface of the grounded sphere so as to maintain the sphere at ground potential. Using the method of images, the real induced surface charge can be simulated by a single imaginary point charge q_i inside the sphere that has the same effect of maintaining a zero isopotential surface on the outside of the sphere.

If such a charge exists then by symmetry it must lie on the line connecting q with the center of the sphere. Suppose that the charge "q" and image charge q_i give zero potential at the points $P1$ and $P2$. It follows from the definition of potential at a point, designated by $\Phi(P)$, that

$$4\pi\varepsilon_0\Phi(P1) = q/(D + a) + q_i/(a + b) = 0$$

and

$$4\pi\varepsilon_0\Phi(P2) = q/(D - a) + q_i/(a - b) = 0$$

hence

$$q_i = -aq/D \qquad \text{and} \qquad b = a^2/D$$

To show that this image charge positioned at b gives zero potential at a general point $P3$:

$$\Phi(P_3) = q/(4\pi\varepsilon_0 r_1) + q_i/(4\pi\varepsilon_0 r_2)$$

now

$$r_1 = (D^2 + a^2 + 2Da\cos\varphi)^{0.5}$$

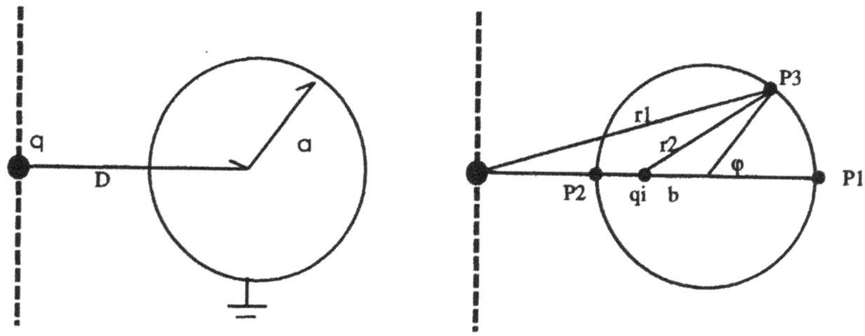

FIGURE C-2.1. Method of images.

and

$$r_2 = (b^2 + a^2 + 2ba \cos \varphi)^{0.5}$$

so

$$4\pi\varepsilon_0 \Phi(P_3) = q(D^2 + a^2 + 2Da \cos \varphi)^{-0.5} - (aq/D)(a^4/D^2 + a^2 + 2a^3 \cos \varphi/D)^{-0.5}$$

Inspection of the second term on the right-hand side shows that it is equal to the first term. The right hand side is therefore zero and hence $\Phi(P_3) = 0$.

C-2.2. *Partial Discharge of Nonconductive Disc to Grounded Sphere*

The analysis can readily be extended to the case of a grounded sphere above the center of a uniformly charged disc which discharges completely up to some arbitrary radius (Figure C-2.2). Consider first a charged annulus on the disc with surface charge density *s*. The disc has radius *S* and the charged region lies between radii *R* and *S*. A charged surface element of area $r\,dr\,d\theta$ therefore has a charge $sr\,dr\,d\theta$ and imposes an image charge dq_i inside the sphere at a distance b_1 from its center, on the line joining the elemental area to the center of the sphere.

From the relationships established previously for magnitude and location of the image charge:

$$b_1 = a^2(D^2 + r^2)^{-0.5}$$

$$dq_i = -as(D^2 + r^2)^{-0.5}\, r\, dr\, d\theta$$

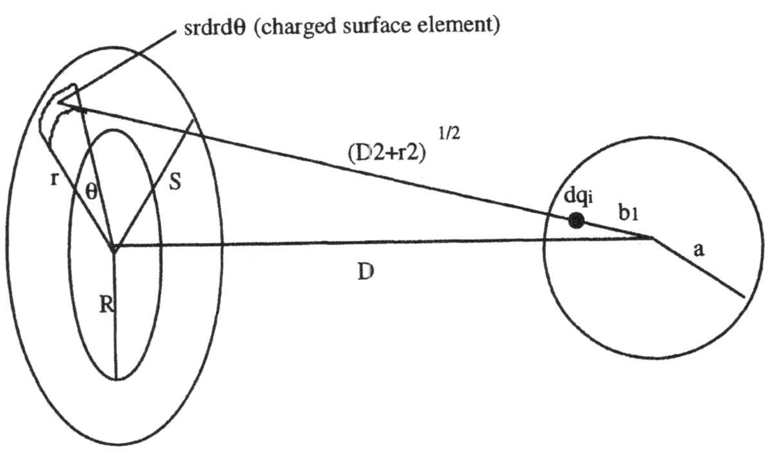

FIGURE C-2.2. Grounded sphere above charged nonconductive disc.

The total induced charge on the sphere due to all elements of the annulus can be found by integration:

$$q_i = \int_0^{2\pi}\int_R^S -as(D^2 + r^2)^{-0.5} r \, dr \, d\theta$$

$$q_i = -2\pi sa \int_R^S r(D^2 + r^2)^{-0.5} dr$$

$$q_i = -2\pi sa\{(D^2 + S^2)^{0.5} - (D^2 - R^2)^{0.5}\}$$

If the disc is uniformly charged over its entire surface the image charge corresponds to the case $R = 0$.

The image charge is not located at a single point but is distributed over a continuous closed surface inside the sphere. Although the induced charge is in reality distributed on the surface of the sphere, its magnitude is identical to that of the image charge. This can be seen by considering a Gaussian surface located just outside the sphere. The total flux through this surface must be the same whether the sphere has an induced surface charge or whether the latter is replaced by an image charge.

C-2.3. *Effect of Image Charge on Detectable Charge Transfer from Disc to Sphere*

A discharge between the charged disc and the sphere may be characterized by measuring the flow of charge, or current, from the sphere to ground. Integration of the current-time measurement gives the charge transferred. In practice the measured charge is reduced by the presence of the induced (image) charge [71].

Before the discharge the entire surface is uniformly charged. Thus the initial image charge q_{i0} can be found as above using $R = 0$

$$q_{i0} = -2\pi sa\{(D^2 + S^2)^{0.5} - D\}$$

This is the total image charge due to a uniformly charged disc and equals the induced charge on the spherical electrode. To get an idea of the magnitude of this induced charge, one can compare it with the total initial charge on the disc ($q_0 = s\pi S^2$) and substitute typical values for geometry. Let an electrode of radius $a = 1$ cm be positioned with its center 10 cm above a charged disc of radius $S = 20$ cm. In this case $D = 10$ cm and from

$$|q_{i0}/q_0| = 2a\{(D^2 + S^2)^{0.5} - D\}/S^2$$

it follows

$$|q_{i0}/q_0| = 6.2\%$$

- The induced charge varies with geometry but is typically of the order 10% or less of the total charge, so errors are not usually large. An exception is discussed in C-2.4.

Suppose all the charge on the disc contained up to inner radius R is discharged. Then, the final image charge q_{if} can be found as above by setting $R = R$:

$$q_{\mathrm{if}} = -2\pi sa\{(D^2 + S^2)^{0.5} - (D^2 + R^2)^{0.5}\}$$

The charge that has been transferred to the sphere q_{t} is

$$q_{\mathrm{t}} = \pi R^2 s$$

When the discharge occurs, suppose a quantity of charge δ flows to ground through the measuring device. We can now balance the charges on the sphere:

$$q_{\mathrm{if}} = q_{\mathrm{i0}} + \pi R^2 s - \delta$$

Hence

$$\delta = 2\pi sa\{D - (D^2 + R^2)^{0.5}\} + \pi R^2 s$$

This equation shows that the measured charge transfer δ is independent of the radius S of the charged disc. The ratio of measured charge transfer to the true charge transfer is given by

$$\delta/q_{\mathrm{t}} = \{R^2 - 2a[(D^2 + R^2)^{0.5} - D]\}/R^2$$

This analysis shows that the measured charge transfer can never be equal to the actual charge transfer except where $q_{\mathrm{if}} = q_{\mathrm{i0}}$, that is, either there is no change in image charge during discharge or the image charge is so small that any change is less than the sensitivity of the measuring device. This can be an important effect for accurate measurement where the geometry makes the image charge a large fraction of the charge transferred and must be considered when measuring charge transfer. In practice the image charge can be made very small compared with q_{t} by using a special probe. For example, a spherical electrode can contain an isolated detection element which receives the total charge transfer from the discharge [198]. The image charge on this element is a small fraction of the total image charge on the probe.

C-2.4. Limit Cases: Discharge from Disc to Grounded Plate or Pointed Electrode

As the radius of the grounded sphere becomes larger the magnitude of the image charge becomes greater and the fraction of measurable charge flowing to ground becomes smaller. In the limiting case, consider a grounded circular plate above the charged disc. The distance between the charged disc and grounded plate $d = (D - a)$. We wish to examine the limit of (δ / q_t) as the sphere diameter becomes infinite.

Now

$$\delta / q_t = \{R^2 - 2a[(D^2 + R^2)^{0.5} - D]\} / R^2$$

Substituting $D = d + a$

As $a \to \infty$, $\delta / q_t \to 0$ and (if the electrode is pointed) $a \to 0$ and $\delta \to q_t$.

- The first result ($a \to \infty$) is most significant regarding the measurement of charge from bulking brush discharges inside silos.

Should this be attempted using an isolated, cylindrical metal detector around the inner silo wall, a substantial fraction of the charge transferred in the discharge will already exist as an image charge on the detector and the current flowing from the isolated section to ground will tend to zero. To avoid this, the detector inside the silo can be subdivided into a series of isolated circumferential strips with the expectation that any single discharge will impact only one such strip while the image charge will be shared with them all.

C-2.5. Field and Potential between Charged Nonconductive Disc and Grounded Sphere

An extension of the approach given above allows electric fields and potentials to be calculated (Figure C-2.5). This approach was outlined in [36]. On the axis of symmetry, the solutions can be obtained analytically.

As before, the magnitude and position of the image charge dQ' due to the element of surface charge dQ is given by

$$dQ' = -(a/p)\, dQ$$

and

$$pp' = a^2$$

We can find the equation governing the position of all the image charges dQ'. In the coordinate system with center at the midpoint of the sphere the position of any image charge is given by

$$(p')^2 = x^2 + y^2 + z^2$$

For the similar triangles

$$p'/x = p/D = a^2/(Dp')$$

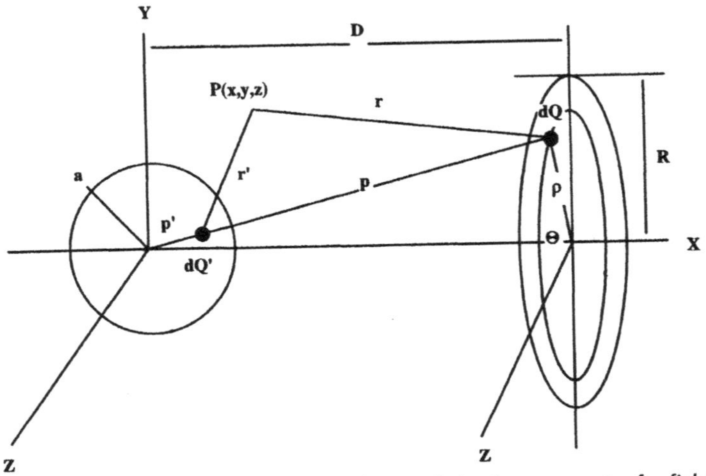

FIGURE C.2.5. Charged nonconductive disc and grounded sphere: geometry for field and potential analysis.

Hence

$$(p')^2 = a^2x/D$$

Thus

$$a^2x/D = x^2 + y^2 + z^2$$

Rearranging and completing the square the location of all image charges dQ' is given by

$$\{x - a^2/(2D)\}^2 + y^2 + z^2 = a^4/(4D^2)$$

Suppose the charge dQ on the disc is due to an element of surface df with charge density s. The contribution $d\Phi$ to total potential at some arbitrary point P at distances r and r', respectively, due to the charge dQ and its image charge dQ' can be written in terms of df:

$$d\Phi = \{sdf/(4\pi\varepsilon_0)\}\{(1/r) + (a/pr')\}$$

In order to find the potential at P due to all the charges dQ and image charges dQ' we need to find the quantities r and r' in terms of the geometrical parameters a, R, and D. Summation is done by integrating over the surface from $\rho = 0$ to R and from $\theta = 0$ to 2π.

From the geometry it follows that

$$r = \{(D - x)^2 + (\rho \sin\theta - y)^2 + (\rho \cos\theta - z)^2\}^{0.5}$$

If we let the coordinates of the image charge dQ be given by x', y', and z', it follows:

$$r' = \{(x-x')^2 + (y-y')^2 + (z-z')^2\}^{0.5}$$

By consideration of the congruent triangles defined by D and p in the x, y, and z planes, and the relative positions of dQ and dQ', it follows that

$$z'/p' = \rho\cos\theta/p \quad \text{and} \quad p'/p = z'/\rho\cos\theta$$

$$y'/p' = \rho\sin\theta/p \quad \text{and} \quad p'/p = y'/\rho\sin\theta$$

$$x'/p' = D/p \quad \text{and} \quad p'/p = x'/D$$

Since, from the earlier discussion of the image charge position, $pp' = a^2$:

$$p'/p = a^2/p^2$$

also

$$p^2 = \rho^2 + D^2$$

hence

$$p'/p = a^2/(\rho^2 + D^2)$$

The potential at the general point $P(x,y,z)$ is therefore given by

$$\Phi(P) = \{s/(4\pi\varepsilon_0)\}\int_0^R\int_0^{2\pi}\{[(D-x)^2 + (\rho\sin\theta - y)^2 + (\rho\cos\theta - z)^2]^{-0.5}$$

$$-(a/p)[(x-x')^2 + (y-y')^2 + (z-z')^2]^{-0.5}\}\rho\cdot d\theta\cdot d\rho$$

C-2.5.1. On-Axis Potential

This integration is too complex for analytical solution but may be simplified by considering only points on the X-axis where $z = y = 0$. The above equations can be used to substitute for x', y' and z'. Hence:

$$\Phi(P)_X = (s/2\varepsilon_0)\int_0^R\{[(D-x)^2 + \rho^2]^{-0.5} - (a/p)[(x-x')^2 + y'^2 + z'^2]\}^{-0.5}\rho\, d\rho$$

$$\Phi(P)_X = (s/2\varepsilon_0)\left\{\left\{[(D-x)^2 + \rho^2]^{0.5}\right\}_0^R - I_2\right\}$$

$$I_2 = \int_0^R (a\rho d\rho)/\{p[(x - Da^2/p^2)^2 + \rho^2 a^4/p^4]^{0.5}\}$$

Expanding the denominator, substituting $p^2 = \rho^2 + D^2$ and integrating

$$I_2 = \left\{(a/x^2)[x^2\rho^2 + (xD - a^2)^2]^{0.5}\right\}_0^R$$

After substitution and rearrangement the solution for axial potential is

$$\Phi(P_X) = (s/2\varepsilon_0)\{(D-x)[\{1 + (R/(D-x))^2\}^{0.5} - 1] - (a/x)\{D - (a^2/x)\}$$

$$\times \{\{1 + (R/(D - a^2/x))^2\}^{0.5} - 1]\}$$

C-2.5.2. On-Axis Electric Field

The field strength is given by $E(P_X) = -\text{grad}\,\Phi(P_X)$

$$E(P_X) = -(s/2\varepsilon_0)\{1 - (D-x)\,[(D-x)^2 + R^2]^{-0.5} - aD/x^2 + 2a^3/x^3 - dT/dx\}$$

where

$$dT/dx = d/dx\,\{(a/x)[(D - a^2/x)^2 + R^2]^{0.5}\}$$

$$dT/dx = -(a/x^2)\{(D - a^2/x)^2 + R^2\}^{0.5} + (a/x)\,(D - a^2/x)(a^2/x^2)\{(D - a^2/x)^2 + R^2\}^{-0.5}$$

$$dT/dx = -(a/x^2)\{(D - a^2/x)^2 + R^2\} + (a/x)\,(D - a^2/x)(a^2/x^2)\{(D - a^2/x)^2 + R^2\}^{-0.5}$$

The solution for axial field strength is therefore given by

$$E(P_X) = -(s/2\varepsilon_0)\,[1 - (D-x)\{(D-x)^2 + R^2\}^{-0.5} - aD/x^2 + 2a^3/x^3 - (a/x^2)\,\{(D - a^2/x)\,(2a^2/x - D) - R^2\}\{(D - a^2/x)^2 + R^2\}^{-0.5}]$$

- For a given surface charge density s the field close to the electrode decreases rapidly with increased distance. If gas breakdown occurs this local drop in field corresponds to the largest energy drop as charge moves toward the electrode.

Observation of brush discharges shows that the brightest part of the discharge occurs as a luminous root or stem close to the electrode and ignition must occur in this region (2-5.1.1). However, over small distances between about 2–5 mm from a large diameter electrode the field is approximately constant (±10–15%). This approximation is used in (2-5.1.1) to estimate the effective energy of a brush discharge.

C-2.5.3. Field Intensification and Maximum Surface Charge Density in Presence of Electrode

Field intensification occurs when a grounded electrode is introduced into the space above a charged surface. The same effect occurs as a field meter head is brought up to a charged surface. To simplify interpretation of the field readings the meter can be equipped with a grounded annulus and brought close to the surface so that a uniform, unidirectional field is produced (C-2.5.5).

The field intensification produced by a spherical electrode can be found from the previous equation. At the surface of the electrode on the X-axis, $x = a$ and the field:

$$E_a = -(s/2\varepsilon_0)[3 - (D/a) + \{(D-a)(D/a - 3) + R^2/a\}\{(D-a)^2 + R^2)^{-0.5}\}]$$

This equation was originally presented by Heidelberg [36].

The undisturbed field above an isolated charged disc (see C-2.5.4) is intensified at the electrode surface by the geometry-dependent factor F:

$$E_a = -(s/2\varepsilon_0)F$$

The former equation can be used to estimate the maximum allowable surface charge density in air by solving the case where the field at the electrode (E_a) equals the breakdown field of air (3 MV/m). For example, consider a 1-cm-radius electrode (a = 0.01 m) positioned 5 cm (d = 0.05 m) above a charged nonconductive disc of large radius (R = 0.5 m). In this case $D = (a + d) = 0.06$ m. For a breakdown field of 3.0 MV/m the maximum allowable charge density s is 1.20×10^{-6} C/m^2. The field intensification factor F in this case is 44. As shown in C-2.5.4, the maximum allowable surface charge density for an isolated disc in air is 5.3×10^{-5} C/m^2. Hence the presence of this electrode reduces the maximum allowable charge density to about 2.3% of the value for an isolated disc.

As shown in C-2.5.5 the maximum surface charge density is 2.65×10^{-5} C/m^2 if a flat, grounded plate is used to produce a uniform, unidirectional field above the charged disc. In the presence of the electrode described above, the disc supports only about 4.5% of the surface charge density possible with a uniform, unidirectional field.

- For equal gaps, smaller diameter electrodes produce greater field intensification and air breakdown occurs at smaller surface charge densities. Brush discharges are therefore more likely to occur at small electrodes although the charge and energy involved is less (2-6.2).

C-2.5.4. Maximum Surface Charge Density for Isolated Charged Disc in Air

It follows from the axial field equation in C-2.5.2 that the (bidirectional) field due to a charged nonconductive disc when there is no electrode or other disturbing influence present ($a = 0$ and $x = D$) is given by

$$E_x = -s/2\varepsilon_0$$

Neglecting attenuation due to the dielectric constant of the nonconductive disc, the same field must also emanate from the other side of the disc. Air has a uniform breakdown field of about 3 MV/m and a permittivity of about 8.85×10^{-12} F/m.

- The maximum surface charge density s that can be supported prior to air breakdown in the case of a uniform, bidirectional field is 5.3×10^{-5} C/m^2.

C-2.5.5. *Use of Field Meter to Measure Surface Charge Density*

Suppose a field meter is used to measure the field above a charged nonconductive disc by bringing it axially toward the center of the disc. The field meter head is assumed to be equipped with a shielding annulus. This is designed to produce an approximately uniform field at the sensing aperture provided the separation from a charged surface is small compared with the diameter of the shielding annulus. To estimate the electric field at the sensing aperture, consider the limit of equation C-2.5.3 as the diameter of the spherical electrode a becomes very large. First replace D with $(d + a)$. In the limit $a \rightarrow \infty$ and the field is

$$E_a \rightarrow -(s/\varepsilon_0)\{1 - 1/[1 + (R/d)^2]\}$$

When R is large compared with separation d, the term $\{1 - 1/[1+(R/d)^2]\}$ approaches unity and $E_a \rightarrow (-s/\varepsilon_0)$. This field is double that produced in the case of an isolated charged disc (C-2.5.4).

- The field meter doubles the field relative to an isolated nonconductive disc, since the field is all in the direction of the grounded body. The maximum surface charge density that can be supported prior to air breakdown in a uniform, unidirectional field is 2.65×10^{-5} C/m^2.

These results have application to the interpretation of field measurements on charged belts, webs and FIBCs. When a shielded field meter is brought away from a charged plastic surface, the observed field reading decreases rapidly when the separation "d" is no longer small compared with the surface width R, owing to the $\{1 - 1/[1 + (R/d)^2]\}$ dependence.

C-2.5.6. *Surface Potential on Charged Disc*

Unlike a conductor, a charged nonconductive disc does not have an isopotential surface but instead the capacitance between the disc and grounded electrode depresses the surface potential as a function of radius across the disc. While instruments are commercially available to "measure" the surface potential of nonconductors such as belts and foils, the potential reading is not quantitative and represents only a proportionality to the surface charge density. For instruments calibrated in volts, calibration is typically done using a metal plate held at fixed potential and there is no equivalence with the potential on a nonconductor. If the surface potential is "measured" using a field meter the value will be severely underestimated relative to a metal plate of constant potential, owing to the potential depression by the induced countercharge on the grounded meter head.

REFERENCES

1. CENELEC (European Commission for Electrotechnical Standardization) "Safety of Machinery—Guidance and Recommendations for the Avoidance of Hazards Due to Static Electricity," CLC/TC44X (Secretariat) 161.
2. BS 5958 "Code of Practice for Control of Undesirable Static Electricity," Part 1 "General Considerations" and Part 2 "Recommendation for Particular Industrial Situations," British Standards Institution, London.
3. API RP2003 "Protection Against Ignitions Arising Out of Static, Lightning and Stray Currents," American Petroleum Institute, Washington DC.
4. ESCIS, Static electricity rules for plant safety, *Plant/Operations Progress*, 7, No. 1, January 1988.
5. "International Safety Guide for Oil Tankers and Terminals, (ISGOTT), Witherby & Co Ltd., London.
6. NPCA Scientific Circular 803 "Generation and Control of Static Electricity, National Paint and Coatings Association Inc., Washington DC.
7. Joint Departments of the Army and Air Force USA, Technical Manual TM 5-811-3/AFM 88-9, Chapter 3, "Electrical Design Lightning and Static Electricity Protection."
8. Britton, L.G., and Smith, J. A., Static hazards of drum filling: Parts 1 and 2, *Plant/Operations Progress*, 7, No. 1, January 1988.
9. Britton, L.G., Using material data in static hazard assessment, *Plant/Operations Progress*, 11, No. 2, April 1992.
10. Britton, L.G., Static hazards using flexible intermediate bulk containers for powder handling, *Process Safety Progress,* 12, No. 4, October 1993.
11. Britton, L.G., Estimating the minimum ignition energy of hybrid mixtures, *Process Safety Progress,* 17, No. 2, Summer 1998.
12. Britton, L.G., Systems for electrostatic evaluation in industrial silos, *Plant/Operations Progress*, 7, No. 1, January 1988.
13. Luttgens, G., Collection of accidents caused by static electricity, *J. Electrostatics,* 16, Nos. 2&3, May 1985.
14. Hughes, J.F., et al., Explosion hazards and diagnostic techniques associated with powder handling in large silos, *Inst. Phys. Conf. Ser.* No 27, pp. 264–275 London, 1975.
15. Blythe, A.R., and Reddish, W., Charges on powders and bulking effects, *Inst. Phys. Conf. Ser.* No. 48, London, 1979.
16. Owens, J., Spark ignition hazards caused by charge induction, *Plant/Operations Progress*, 7, No. 1, January 1988.

17. Britton, L.G., and Kirby, D.C., Analysis of a dust deflagration, *Plant/Operations Progress*, 8, No. 3, July 1989.
18. Bailey, A.G., Electrostatic phenomena during powder handling, *Powder Technology* 37, 71–85, 1984.
19. Gibson, N., Antistatic valves in British Standard, *J. Electrostatics*, 1, 315, 1975.
20. Gibson, N., Effect of electrification in ball valves on ignition risk in liquid pipeline systems, *J. Electrostatics* 1, 339–350, 1975.
21. Bulkley, W.L., "Static Electricity Effects in Tank Truck Loading." Presented at 1970 National Safety Congress, Petroleum Section, October 28th, 1970.
22. Strawson, H., and Lyle, A.R., Safe charge densities for road and rail tank car filling, *Inst. Phys. Conf. Ser.* No. 27, Chapter 4, 1975.
23. Walmsley, H.L., The avoidance of electrostatic hazards in the petroleum industry, *J. Electrostatics*, 27, Nos. 1–2, 1992.
24. Bustin, W.M., and Dukek, W.G., *Electrostatic Hazards in the Petroleum Industry,* Research Studies Press Ltd., Letchworth, England, 1983.
25. Glor, M., *Electrostatic Hazards in Powder Handling,* Wiley, New York, 1988.
26. Gibson, N., and Harper, D.J., Parameters for assessing electrostatic risk from nonconductors—a discussion, *J. Electrostatics*, 21, 27–36, 1988.
27. Newman, M.N., and Robb, J.D., "Investigation of Minimum Corona Type Currents for Ignition of Aircraft Fuel Vapors," NASA Technical Note D-440, Langley Field, VA, 1960.
28. Sloane, R.W., The ignition of gaseous mixtures by the corona discharge, *Phil. Mag.*, 998–1011, 1935.
29. Maurer, B., Glor, M., Luttgens, G., and Post, L., Hazards associated with propagating brush discharges on flexible intermediate bulk containers, compounds and coated materials, *Inst. Phys. Conf. Ser.* No. 85, Section 3, Oxford, 1987.
30. Pratt, T.H., et al., "Electrostatic Hazards in Liquid Pumping Operation," Paper 57-1, 6th Int. Symp. Loss Prevention and Safety Promotion in the Process Industries, Oslo, June 1989.
31. (Anon), "Static Spark Flashes 'Empty' Styrene Drum," MCA Case History No. 41.
32. Thorpe, D.G.L., et al., Electrostatic Hazards in Sugar Dust in Storage Silos, *J. Electrostatics*, 16, 193–207, 1985.
33. Pratt, T.H., "Electrostatic Ignitions of Fires and Explosions," Burgoyne Inc., 1997 (available through AIChE-CCPS).
34. Schwenzfeuer, K., and Glor, M., Tests to determine the ignition of dust by brush discharges, *J. Electrostatics*, 30, 115–122, 1993.
35. Wright, L., and Ginsburgh, I., What experimentation shows about static electricity, in Vervalin (Ed.), *Fire Protection Manual for Hydrocarbon Processing Plants,* Gulf Publishing Co, Houston, p. 235, 1973.
36. Heidelberg, E., "Generation of Igniting Brush Discharges by Charged Layers on Earthed Conductors," 1967 Static Electrification Conference, Institute of Physics, London.
37. Heidelberg, E., *Ignition of Explosive Atmospheres by Static Electricity,* PTB (Physikalisch-Technische Bundesanstalt), Mitteilungen, Germany, 1970.
38. Wilson, N., The ignition of natural gas by spark discharges from the body, *Inst. Phys. Conf. Ser.* No. 66, pp. 21–27, Institute of Physics, London, 1983.
39. Tabata, Y., et al., Safety control of incendiary discharge occurring from active electrostatic eliminator, *J. Electrostatics*, 30, 93–102, 1993.
40. Durkin, W.J., Dangers in interpreting measurements on plastic webs, *J. Electrostatics*, 35, 215–229, 1995.
41. Taillet, J., et al, Testing a new device for suppressing electrostatic charges on polymer granules, *J. Loss Prev. Process Ind.*, 5, No. 4, 235–237, 1992.

42. Ballal, D.R., and Lefebvre, A.H., The influence of spark discharge characteristics on minimum ignition energy in flowing gases, *Combustion and Flame*, 24, 99–108, 1975.
43. Neer, M.E., Autoignition of flowing hydrogen–air mixtures, *AIAA Journal*, 13, No. 7, 924–928, 1975.
44. Singh, S., A study of electrostatic activity in a full scale pneumatic conveying system, *J. Powder & Bulk Solids Technology*, 7, No. 2, 13–16, 1983.
45. Tolson, P., Examination of possible electrostatic hazards arising from the use of water spray barriers to disperse flammable vapors, *J. Electrostatics*, 23, 89–98, 1989.
46. Seifert, H., et al., "Steam Curtains—Effectiveness and Electrostatic Hazards," 4th Int. Symp. Loss Prevention & Safety Promotion in the Process Industries, Harrogate, England, Sept.12–16, 1983.
47. Pratt, T.H., Electrostatic considerations for automatic tank gauges, *J. Loss Prev. Process. Ind.*, 5, No. 2, 1992.
48. Pischinger, S., and Heywood, J.B., "How Heat Losses to the Spark Plug Electrodes Affect Flame Kernel Development in a SI-Engine," SAE Technical Paper 900021, SAE Int. Congress & Expo, Detroit, MI, Feb. 26–Mar 2, 1990.
49. Kramer, H., and Frochtenigt, H., Electrostatic charging of spray cans, *J. Electrostatics*, 30, 159–164, 1993.
50. Glor, M., et al., Recent Developments in the assessment of electrostatic hazards associated with powder handling, *Loss Prev. and Safety Promotion in the Process Industries*, 1, 221, 1995.
51. Chubb, J.N., and Butterworth, G.J., Instrumentation and techniques for monitoring and assessing electrostatic ignition hazards, *Inst. Phys. Conf. Ser.* No. 48, p. 88, 1979.
52. Butterworth, G.J., Electrostatic ignition hazards associated with the preventative release of fire extinguishing fluids, *Inst. Phys. Conf. Ser.* No. 48, pp. 161–169, 1979.
53. Wilson, N., The nature and incendiary behavior of spark discharges from the body, *Inst. Phys. Conf. Ser.* No. 48, p. 74, 1979.
54. "Assessing Flame Resistant Clothing Use," CMA Managers Guide, Chemical Manufacturers Assoc., July 1997.
55. Klinkenberg, A., and van der Minne, J.L., *Electrostatics in the Petroleum Industry*, p. 22, Elsevier, New York, 1958.
56. Blanc, M.V., et al., Ignition of explosive gas mixtures by electric sparks, *3rd Symp. on Combustion and Flame and Explosion Phenomena*, p. 365, Williams and Wilkins Co., 1949.
57. Glor, J., Ignition of gas/air mixtures by discharges between electrostatically charged plastic surfaces and metallic electrodes, *J. Electrostatics*, 10, 327–332, 1981.
58. Lovstrand, K.G., The ignition power of brush discharges—Experimental work on the critical charge density, *J. Electrostatics*, 10, 161–168, 1981.
59. Bartknecht, W., *Dust Explosions*, pp. 106–107, Springer-Verlag, New York, 1989.
60. Schwenzfeuer, K., and Glor, M., "The Safety Concerns between Conductive Type C and Static Dissipating FIBCs," Institute of Safety and Security, Switzerland, 16th January 1998.
61. Dorsey, J.S., Static sparks—How to exorcise go devils, *Chemical Engineering*, September 13, 1976.
62. Movilliat, P., and Giltaire, M., Mesure de l'energie d'inflammation de melanges gazeux par decharge capacitive: inflammation par decharge d'une personne chargee d'electricite statique, *J. Electrostatics*, 6, 307–331, 1979.
63. Magison, E.C., *Electrical Instruments in Hazardous Locations*, 3rd ed., ISA, 1978.
64. Ballal, D.R., and Lefebvre, A.H., Ignition and quenching of quiescent fuel mists, *Proc. Roy. Soc. Lond. A*, 364, 227–294, 1978.
65. Spalding, D.B., *Combustion and Mass Transfer*, Pergamon Press, Oxford, 1979.

66. Peters, J.E., and Mellor, A.M., An ignition model for quiescent fuel sprays, *Combustion and Flame*, 38, 65–74, 1980.
67. Luttgens, G., and Glor, M., *Understanding and Controlling Static Electricity,* Expert Verlag, 1989.
68. Kramer, H., and Asano, K., Incendivity of sparks from electrostatically charged Liquids, *J. Electrostatics*, 6, No. 4, 1979.
69. Carruthers, J.A., and Wigley, K.J., The estimation of electrostatic potentials, fields and energies in a rectangular metal tank containing charged fuel, *J. Inst. Petroleum,* 48, No. 462, 180–195, 1963.
70. Asano, K., Electrostatic potential and field in a cylindrical tank containing charged liquid, *Proc. IEEE,* 124, No. 12, 1977.
71. Britton, L.G., and Williams, T.J., Some characteristics of liquid-to-metal discharges involving a charged 'low risk' oil, *J. Electrostatics*, 13, No. 2, 1982.
72. Haase, H., *Electrostatic Hazards: Their Evaluation and Control,* Verlag Chemie, New York, 1977.
73. Jones, T.B., and King, J.L., *Powder Handling and Electrostatics: Understanding and Preventing Hazards,* Lewis Publishers, Chelsea, MI, 1992.
74. Anon., "Interim Report. Tanker Accident Study Committee," API, Washington DC, November 1970.
75. Lange, R.F., "Shoreside Static Electrification Studies during Tank Washing in a 25 Foot Diameter Tank," Mobil Research and Development Corp., Paulsboro, NJ, Report No. 70.33-AD, November 13, 1970.
76. Gibson, N., "Static Electricity in Fluids," Paper 9, Institute of Physics Conference Series Number 11, London, 1971.
77. Schon, G., and Kramer, H., "On the Size of Stationary Space Clouds in Streaming Media," Paper 14, Institute of Physics Conference Series Number 11, London, 1971.
78. van der Meer, D., "Electrostatic Charge Generation During Washing of Tanks with Water Sprays I: General Introduction," Paper 15(i), Institute of Physics Conference Series Number 11, London, 1971.
79. van de Weerd, J.M., "Electrostatic Charge Generation During Washing of Tanks with Water Sprays II: Measurements and Interpretation," Paper 15(ii), Institute of Physics Conference Series Number 11, London, 1971.
80. Smit, W., "Electrostatic Charge Generation during the Washing of Tanks with Water Sprays III: Mathematical Methods," Paper 15(iii), Institute of Physics Conference Series Number 11, London, 1971.
81. Vos, B., "Electrostatic Charge Generation during the Washing of Tanks with Water Sprays IV: Mechanism Studies," Paper 15(iv), Institute of Physics Conference Series Number 11, London, 1971.
82. Klaver, R.F., "Electrical Field Strength in Cargo Tanks of Seven Tankers during or after Washing," Chevron Research Co., Richmond CA, November 2, 1971.
83. Anon., "Interim Report, Tanker Accident Study Committee," API, Washington DC, November 31, 1971.
84. Bustin, W.M., "Electrically Charged Water Mist Produced by Water Washing," ESSO Engineering Technology Dept., Report No. EE.8TMR.72, March 1972.
85. Barreto, E., "The Oil Tanker and the Charged Mist," Special Meeting of Experts, Electrostatics in Tank Washing, Sponsored by Int. Chamber of Shipping, Snekkersten, Denmark, 17-19 April 1972.
86. van de Weerd, J.M., "Can Probe-to-Cloud Discharges be an Ignition Source for Tanker Explosions?" Shell Research NV, The Hague, Report No. AMSR.0016.72, 1972.

87. Napier, D.H., "Generation of Static Electricity in Steam Screens and Water Curtains," Paper P, Institution of Chemical Engineers Symposium Series No. 39, London,, 1974.
88. Klaver, R.F., "Final Report: Charge Transfer between Isolated Conducting Objects and Ground in a Tank Filled with an Electrically Charged Mist," Chevron Research Co. Report for API Statics Research Program, May 28, 1974.
89. Hughes, J.F., et al., "Incendive Sparks Originating from Charged Water Mists," Nat. Acad. Sci., Annual Conf. on Electrical Insulation and Dielectric Phenomena, Washington DC, pp. 33–42, 1974.
90. Anon., "Final Report, Tanker Accident Study Committee," API, Washington DC, November 27, 1974.
91. Bright, A.W., et al, Research on electrostatic hazards associated with tank washing in very large crude carriers (supertankers): I. Introduction and experimental modelling, *J. Electrostatics,* 1, No. 1, February 1975.
92. Bassett, J.D., Research on electrostatic hazards associated with tank washing in very large crude carriers (supertankers): II. Study of the charging mechanism during the production of water aerosols, *J. Electrostatics,* 1, No. 1, February 1975.
93. Chubb, J.N., Practical and computer assessments of ignition hazards during tank washing and during wave action in part-ballasted OBO cargo tanks, *J. Electrostatics,* 1, No. 1, February 1975.
94. van der Meer, D., Electrostatic charge generation during tank washing: Project summary and introduction, *J. Electrostatics,* 1, No. 3, August 1975.
95. Lindbauer, R.L., Electrostatic charge generation during tank washing: Reducing hazardous space potentials in tankers: A Theoretical and experimental approach, *J. Electrostatics,* 1, No. 3, August 1975.
96. van de Weerd, J.M., Electrostatic charge generation during tank washing: Spark mechanisms in tanks filled with charged mist, *J. Electrostatics,* 1, No. 3, August 1975.
97. Butterworth, G.J., et al., "Studies of Tank Washing on *British Purpose* during Ballast Voyage from Isle of Grain to Cape of Good Hope April 7th to May 5th 1977," Electrostatics and Physics Group, UKAEA Culham Laboratory, Report No. CLM/RR/D2/34, February 1978.
98. Butterworth, G.J., The detection and characterisation of electrostatic sparks by radio methods, *Inst. Phys. Conf. Ser.* No. 48, pp. 97–105, London, 1979.
99. Hughes, J.F., Electrostatic charges on board ships, *Trans. I. Mar. E.* 92, TM6, 1980.
100. Makin, B., and Lees, P., Measurement of charge transfer in electrostatic discharge, *J. Electrostatics,* 10, May 1981.
101. Butterworth, G.J., et al., Assessment of electrostatic ignition hazards during tank filling with the aid of computer modelling, *J. Electrostatics,* 13, No. 1, August 1982.
102. Mills, J.S., and Haighton, E.J., Prevention of electrostatic hazards associated with shipboard inert gas operations, *Inst. Phys. Conf. Ser.* No. 66, London, 1983.
103. Kitagawa, T., Presumption of causes of explosions in very large tankers, *Gefahrliche Lading (Dangerous Cargo),* No. 10, October 1975.
104. Jones, M.R.O., and Bond, J., Electrostatic hazards associated with marine chemical tanker operations: Criteria of incendivity in tank cleaning operations, *Chem. Eng. Res. Des.*, 62, September 1984.
105. Jones, M.R.O., and Bond, J., Electrostatic hazards associated with marine chemical tanker operations: Criteria of safety in tank cleaning operations, *Chem. Eng. Res. Des.,* 63, November 1985.
106. Bulgin, D., Static electricity on rubber tyred vehicles, *Brit. J. Applied Phys.* Supplement No. 2, S83–S87.

107. Mancini, R.A., The use (and misuse) of bonding for control of static ignition hazards, *Plant/Operations Progress*, 7, No. 1, January 1988.

108. Bulkley, W.L., Static danger sparks research on accidents, *Modern Bulk Transporter*, pp.18-19, Jan, 1971.

109. Moore, A.D. (Ed.), *Electrostatics and Its Applications*, Wiley, New York, 1973.

110. Cross, J.A., *Electrostatics: Principles, Problems and Applications,* Adam Hilger, Bristol, 1987.

111. Beach, R., Preventing static electricity fires—III, *Chemical Engineering,* February 1, 1965.

112. Dukek, W.G., et al., "Static Electricity Hazards in Aircraft Fuel Systems," Air Force Aero Propulsion Laboratory Report AFAPL-TR-78-56, August 1978.

113. Leonard, J.T., and Affens, W.A., Charging of jet fuel on polyurethane foams, *Inst. Phys. Conf. Ser.* No. 48, London, 1979.

114. Grenich, A.F., and Tolle, F.F., "Electrostatic Safety with Explosion Suppressant Foams," Air Force Wright Aeronautical Laboratories Report AFWAL-TR-83-2015, March 1983.

115. Bird, J.W., A "Comparison of White Reticulated Foam to Rigid Foam as Fire Protection for Aircraft Dry Bays," Air Force Wright Aeronautical Laboratories Report AFWAL-TM-84-237-FIESE, December 1984.

116. Dixon, T., "Gross Voided Flame Arresters for Fuel Tank Explosion Protection," Boeing Aerospace Co Technical Report AFAPL-TR-73-124, February 1974.

117. Military Specification MIL-F-87260 (USAF), "Foam Material, Explosion Suppression, Inherently Electrically Conductive, for Aircraft Fuel Tank and Dry Bay Areas," February 7, 1992.

118. Society of Automotive Engineers, Inc., "Reticulated Polyurethane Foam Explosion Suppression Material for Fuel Systems and Dry Bays," SAE : AIR4170, December 1991.

119. Miaoulis, I.N., et al., Theory for electric charging in flow of low conductivity liquids through screens, *J. Electrostatics*, 25, 287–294, 1990.

120. Miaoulis, I.N., et al., Electric charging in flow of low-conductivity liquids through screens: A comparison of theory and experiments, *J. Electrostatics*, 25, 295–307, 1990.

121. Maurer, B., Discharges due to electrostatic charging of particles in large storage silos, *Ger. Chem. Eng.* No. 4, 189–195, 1979.

122. Wagner, J.F., Modelling of electrostatic phenomena, *AIChE Loss Prevention,* 9, 119–126, 1975.

123. Huber P.W., and Sonin, A.A., A model for the electric charging process in fuel filtration, *AIChE Loss Prevention*, 9, 127–129, 1975. [Note: this paper is a condensed version of 124.]

124. Huber P.W., and Sonin, A.A., A model for the electric charging process in fuel filtration, *J. Colloid and Interface Sci.*, 52, No. 1, September 1975.

125. Huber P.W., and Sonin, A.A., Comments on Gavis and Wagner's correlation for the charging current generated during liquid hydrocarbon filtration, *Chem. Eng. Sci.* 32, No. 3, 331–333, 1977.

126. Lauer, J., and Antal, P.A., Electrostatic charges and flow of organic liquids, *AIChE Loss Prevention*, 6, 34–3, 1972.

127. Post, L., et al., The Avoidance of ignition hazards due to electrostatic charges occurring during the spraying of liquids under high pressure, *J. Electrostatics*, 23, 99–109, 1989.

128. Ballal, D.R., Ignition and flame quenching of quiescent dust clouds of solid fuels, *Proc. Roy. Soc. Lond. A*, 369, 479–500, 1980.

129. Ballal, D.R., Further studies on the ignition and flame quenching of quiescent dust clouds, *Proc. Roy. Soc. Lond. A*, 385, 1–19, 1983.

130. (Anon), "Grounding and Bonding Applications for Control of Static Electricity," Stewart R. Browne Manufacturing Co. Inc., Atlanta, 1993.

131. Zabetakis, M.G., *Safety with Cryogenic Fluids,* pp. 27–28, Plenum Press, New York, 1967.

132. ASTM G 145-96 "Standard Guide for Studying Fire Incidents in Oxygen Systems."

133. Harrop, P.J., *Dielectrics,* Butterworths, London, 1972.

134. Mason, P.I., "Electrostatic Hazards in Liquid Transport," 5th Electrostatics Industrial Course, University of Southampton, September 1975.

135. Bodurtha, F.T., *Industrial Explosion Prevention and Protection,* pp. 41–42, McGraw-Hill, New York, 1980.

136. Secker, P.E., and Chubb, J.N., Instrumentation for electrostatic measurements, *J. Electrostatics*, 16, 1–19, 1984.

137. Meek, J.M., and Craggs, J.D., *Electrical Breakdown in Gases,* Clarendon Press, Oxford, 1953.

138. Loeb, L.B., *Fundamental Processes of Electrical Discharges in Gases,* Wiley, New York, 1939.

139. Kuffel, E., and Abdullah, M., *High Voltage Engineering,* Pergamon Press, New York, 1979.

140. Mackeown, S.S., and Wouk, V., Electrical charges produced by flowing gasoline, *Ind. Eng. Chem.* 34, No. 6, 1942.

141. Bruinzeel, C., Electric discharges during simulated aircraft fueling, *J. Inst. Pet.,* 49, No. 473, May 1963.

142. Lewis, B., and von Elbe, G., Combustion, *Flames and Explosions of Gases,* 2nd ed., p. 333, Academic Press Inc., New York, 1961. Also, 3rd ed., 1987.

143. ASTM D 1125 "Standard Test Method for Electrical Conductivity and Resistivity of Water."

144. ASTM D 2624 "Standard Test Method for Electrical Conductivity of Aviation and Distillate Fuels."

145. ASTM D 3114 "Standard Test Method for D-C Electrical Conductivity of Hydrocarbon Fuels."

146. ASTM D 4308 "Standard Test Method for Electrical Conductivity of Liquid Hydrocarbons by Precision Meter."

147. Goodger, E.M., Light distillate fuels for transport, *J. Institute of Energy,* 68, No. 477, Dec, 1995.

148. Eichel, F.G., Electrostatics, *Chemical Engineering*, pp. 153–167, March 13, 1967.

149. Crowl, D.A., and Louvar, J.F., *Chemical Process Safety: Fundamentals with Applications*, pp. 201-204, Prentice Hall, Englewood Cliffs, New Jersey, 1990.

150. Gibson, N., and Lloyd, F.C., Effect of contamination on the electrification of toluene flowing in metal pipes, *Chem. Engng. Sci.,* 25, 87–95, 1970.

151. Leonard, J.T., and Bogardus, H.F., "Pro-Static Agents in Jet Fuels," NRL Report 8021, Naval Research Laboratory, Washington DC, August 16, 1976.

152. Bruderer, R.E., Use bulk bags cautiously, *Chem. Engng. Progress*, 89, No. 5, 28–31, 1989.

153. Anon, *Electrometer Measurements,* 2nd Ed, Keithley Instruments, Inc., Cleveland, 1977.

154. Greason, W.D., "Methodology to Study the Resistance of Spark Discharges," IEEE Industry Applications Society Annual Meeting, New Orleans, October 5–9, 1997.

155. Pude, J.R.G., and Hughes, J.F., "Spark Energy Measurements Using an Optical Technique," IEEE Industry Applications Society Annual Meeting, San Francisco, October 4–7, pp. 1201–1206, 1982.

156. Larson, R.J., "Vapor Accumulations within Covered Floating Roof Tanks," Annual Meeting of API Codes and Standards Committee, Los Angeles, CA, May 4, 1966.

157. AIChE-CCPS *Guidelines for Engineering Design for Process Safety,* Chapter 11, American Institute of Chemical Engineers, New York, 1993.

158. Johnson, O.W., The Hazard from static electricity on moving rubber-tired vehicles, *Fire Journal*, pp. 25–27, January 1967.

159. Maurer, B., Damage to Enamel on stirred apparatus triggered by static electricity discharges of material being stirred, *J. Electrostatics*, 40&41, 517–522, 1997.

160. Glor, M., Luttgens, G., Maurer, B., and Post, L., Discharges from bulked polymeric granules during the filling of silos—Characterization by measurements and influencing factors, *J. Electrostatics*, 23, 35–43, 1989.

161. Glor, M., and Maurer, B., Ignition tests with discharges from bulked polymeric granules in silos (cone discharges), *J. Electrostatics*, 30, 123–134, 1993.

162. Davie, F.M., et al., Case histories of incidents in heated bitumen storage tanks, *J. Loss Prev. Process Ind.*, 7, No. 3, 217–221, 1994.

163. Whitmore, M.W., et al., Report of an Explosion during the manufacture of an azodicarbonamide formulation, *J. Loss Prev. Process Ind.*, 6, No. 3, 169–175, 1993.

164. Abbey, J.C., and Upham, T.E., An investigation of electrostatically induced failures in Teflon hose, *The Flying A,* 14, No. 4, July 1963.

165. Wood, J.P., Spark hazard ko'd with conductive tubing, *Chemical Processing,* January 1969.

166. Anon., "Ignition of Toluene after Truck Loading," Willcox North America Inc. Incident Report No. 7, 1988.

167. Nanevicz, J.E., "Static Electricity Phenomena : Theory and Problems," SRI International Report, Menlo Park, CA.

168. Britton, L.G., Operating atmospheric vent collection headers using methane gas enrichment, *Process Safety Progress,* 15, No. 4, 1996.

169. Coward, H.F., and Jones, G.W., Limits of flammability of gases and vapors, *US Bureau of Mines Bulletin* 503, 1952.

170. Zabetakis, M.G., Flammability characteristics of combustible gases and vapors, *US Bureau of Mines Bulletin* 627, 1965.

171. Kuchta, J.M., Investigation of fire and explosion accidents in the chemical, mining and fuel-related industries—A manual, *US Bureau of Mines Bulletin* 680, 1985.

172. Harrison, R., Thiele, C., and Hoff, D., "Electrical Hazard Protection of Tank Vessels while Moored to Shore Facilities," JPL Publication 5030-503, Prepared for USDOT-USCG, pp.3-3 and 3-4, May 15, 1981.

173. Bright, A.W., and Britton, L.G., "Study of Hazards in Fuel Tanks of GRP Construction," Southampton University Applied Electrostatics Group, Final Report Ref No 7807/087 to Ministry of Defence Procurement Executive, February 1979.

174. Lees, P., McAllister, D., Smith, J.R., Britton, L.G., and Hughes, J.F., An experimental and computational investigation of electrostatic fields in plastic tanks during filling with low conductivity fuel, *J. Electrostatics*, 10, 267–273, 1981.

175. Saletan, D.I., How to calculate and combat static electricity hazards in process plants II, *Chemical Engineering,* pp. 101–106, June 1959.

176. Robb, J.D., "Energy Storage Mechanisms of Electrostatic Charge in Fuel Tanks," OYEZ-IBC Symposium on Electrostatic Hazards in the Storage and Handling of Liquids and Powders.

177. Crouch, C.A., Conveying: How dilute phase stacks up against dense phase, *Chem. Eng. Prog.*, August 1998.

178. Enstad, G.G., "Electrostatic Charging of Powders by Pneumatic Conveying," Chr. Michelsen Inst. Report Ref. CMI no. 78001-1, April 14, 1978.

179. Williams, A., *Combustion of Sprays of Liquid Fuels,* Elek Science, London, 1976.

180. Green, H.L., and Lane, W.R., *Particulate Clouds: Dusts, Smokes and Mists,* E.&F.N. Spon Ltd., London, 1964.

181. Edwards, H.R., and Underwood, M.C., Ignition of powder/air mixtures by discharges of static electricity, *J. Electrostatics*, 15, 123-115, 1984.

182. *CRC Handbook of Chemistry and Physics*, 58th ed., 1977–1979.

183. Dummer, G.W.A., *Fixed Capacitors,* 2nd ed., Pitman, London, 1964.

184. Cashdollar, K.L., and Chatrathi, K., Minimum explosible dust concentrations measured in 20 liter and 1 cubic meter chambers, *Comb. Sci. and Technol.*, 87, 157–171, 1992.

185. Britton, L.G., Loss case histories in pressurized ethylene systems, *Process Safety Progress,* 13, No. 3, July 1994.

186. Britton, L.G., "A Study of the Electrically Charged Mist Produced during Barge Washing," Union Carbide Project Report, August 1982.

187. Takahashi, Y., Two hundred years of Lichtenberg Figures, *J. Electrostatics*, 6, 1–13, 1979.

188. Moore, W.J., *Physical Chemistry,* 4th ed., p. 838, Longman, New York, 1970.

189. Shell Safety Committee, "Static Electricity—Technical and Safety Aspects," Shell International Petroleum Maatschappij B.V., The Hague, 1988.

190. Walmsley, H.L., Threshold potentials and discharge charge transfers for the evaluation of electrostatic hazards in road tanker loading, *J. Electrostatics*, 26, No 2, 1991.

191. Walmsley, H.L., The calculation of the electrostatic potentials that occur when tanks are filled with charged liquids, *J. Electrostatics*, 26, No. 3, 1991.

192. Walmsley, H.L., Electrostatic ignition hazards in road tanker loading: Part 1. Review and experimental measurements, *J. Electrostatics*, 28, No. 1, 1992.

193. Walmsley, H.L., Electrostatic Ignition Hazards in Road Tanker Loading: Part 2. Statistical analysis for standard conditions, *J. Electrostatics*, 28, No. 2, 1992.

194. Walmsley, H.L., Electrostatic ignition hazards in road tanker loading: Part 3. The variation of risk with loading conditions, *J. Electrostatics*, 28, No. 2, 1992.

195. Walmsley, H.L., The electrostatic potentials generated by loading multiple batches of product into a road tanker compartment, *J. Electrostatics*, 38, No. 3, 1996.

196. Ciba-Geigy Central Safety Committee, "Static Electricity—Rules for Plant Safety," 1978.

197. Herdan, G., *Small Particle Statistics*, Butterworths, London, 1960.

198. Chubb, J.N., and Butterworth, G.J., Charge transfer and current flow measurements in electrostatic discharges, *J. Electrostatics*, 13, No. 2, 1982.

199. Jaeger, N., and Siwek, R., Determination, prevention and mitigation of potential hazards due to the handling of powders during transportation, charging, discharging and storage, *Process Safety Progress,* 17, No. 1, 1998.

200. Zaky, A.A., and Hawley, R., *Conduction and Breakdown in Mineral Oil*, Peregrinus Ltd., London, 1973.

201. Whitaker, G., "The Use of Conductive Plastic Containers in the Handling and Storage of Flammable Solvents and Powders," IBC Conference on Recent Developments in the

Assessment of Electrostatic Hazards in Industry, Tower Thistle Hotel, London, September 1989.

202. Schon, G., *Handbuch der Raumexplosionen* (H.H. Freytag, Ed.) Verlag Chemie, Weinheim Bergstr., 1965.

203. Gibson, N., and Lloyd, F.C., Electrification of toluene flowing in large diameter metal pipes, *J. Appl. Phys.*, 3, 563–573, 1970.

204. Koszman, I., and Gavis, J., Development of charge in low conductivity liquids flowing past surfaces: experimental verification and application of the theory developed for tube flow, *Chem. Eng. Sci.*, 1062, 17, 1023–1040, 1962.

205. Shafer, M.R., Baker, D.W., and Benson, K.R., Electric currents and potentials resulting from the flow of charged liquid hydrocarbons through short pipes, *Journal of Research of the National Bureau of Standards-C. Engineering and Instrumentation*. 69C, No. 4, 307–317, 1965.

206. Britton, L.G., and Hughes, J.F., "Principles for Minimizing Electrostatic Hazards in Fuel Systems," Southampton University Applied Electrostatics Group, Progress Report No. 3 to Ministry of Defence Procurement Executive, Contract NCS 361/75737, June 30, 1980.

207. Bustin, W.M., "Static Electricity Precautions for Plastic Tank Truck Compartments," Exxon Engineering Technical Report EE.56E.84, July 1984.

208. De Soete, G.G., "The Influence of Isotropic Turbulence on the Critical Ignition Energy," 13th Symposium (International) on Combustion, The Combustion Institute, pp. 735–743, 1971.

209. Rogers, G.F.C., and Mayhew, Y.R., *Engineering Thermodynamics*, 2nd ed., p. 493, Longman, New York, 1967.

210. Hay, D.M., and Napier, D.H., Minimum ignition energy of dust suspensions, *I.Chem. E. Symposium Series* No. 49, 73–81, 1977.

211. Ebadat, V., Influence of powder streaming current on surface potential of bulked high density polyethylene, *Inst. Phys. Conf. Ser.* No. 118, *Electrostatics '91*, Oxford, 1991.

212. Britton, L.G., Combustion hazards of silane and its chlorides, *Plant/Operations Progress*, 9, No. 1, 22, 1990.

213. Walmsley, H.L., Electrostatic Hazards from water slugs formed during the washing of ships' tanks: Spark energy calculations, *J. Phys. D: Appl. Phys.* 20, 329–339, 1987.

214. Jones, R.T., Williams, T.J., and Abu Sharkh, S., Assessment of industrial electrostatic hazards using finite element electric field analysis, *J. Electrostatics*, 40&41, 449–454, 1997.

215. Williams, T.J., and Jones, R.T., "Modelling the Electrostatic Ignition Hazards Associated with the Cleaning of Tanks Containing Flammable Atmospheres," Electrostatics '99, Cambridge, March 1999.

216. Dwight, H.B., Calculation of resistances to ground, *Electrical Engineering*, pp. 1319–1328, 1936.

217. Chubb, J.N., Developments in electrostatic fieldmeter instrumentation, *J. Electrostatics*, 14, 349–358, 1983.

218. Chubb, J.N., Instrumentation and standards for testing static control materials, *IEEE Transactions on Industry Applications*, 26, No. 6 November–December 1990.

219. Skinner, B.F., Education in 1984, *New Scientist*, p. 484, May 21, 1964.

220. Finke, J., Electrostatic effects of charged steam jets, *J. Electrostatics*, 23, 69–78, 1989.

221. Strawson, H., Avoiding Electrostatic Hazards while Loading Road and Rail Tank Vehicles with Petroleum Products, *Gemeinschaftstagung OGEW/DGMK* vom 4, bis 6, pp. 630-647, Salzburg, 1976.

222. Rose, H.E., Useful Approach to dust explosion hazards, *AIChE Loss Prevention,* 9, 82–90, 1975.

223. Britton, L.G., "New Compilation of Fundamental Burning Velocities of Fuels in Air," Union Carbide Corporation Memorandum, File No. 33529, February 27, 1985. Subsequently published in NFPA 68.

224. Strehlow, R.A., *Combustion Fundamentals,* p. 383, McGraw-Hill, New York, 1984.

225. Acevedo, J., String makes static control a snap, *Paper, Film and Foil Converter,* April 1998.

226. Private Communication from Graham Hearn, Wolfson Electrostatics, University of Southampton, January–February 1999.

227. Fredholm, O., and Lovstrand, K.G., On the discharge of static electricity from an insulator surface, *J. Phys. E: Scientific Instruments,* 5, 1972.

228. Blythe, A.R., and Carr, G.E., Characteristics of propagating brush discharges on dielectric films, *J. Electrostatics,* 10, 321–326, 1981.

GLOSSARY

Air. Sea level concentrations of the six principal constituents of dry air are given in the following table. Dry air is often obtained by drying compressed atmospheric air using a suitable drying agent. In some cases "dry air" is made up in cylinders using 20.95 mol% oxygen with the balance being exclusively nitrogen. Care should be taken that argon is not analyzed as oxygen, as can occur in GC analysis. Dry air contains a higher concentration of oxygen than atmospheric air, which contains moisture. The absolute humidity of saturated air is found using steam tables. For example, saturated air at 100°F contains 6.46 mol% water vapor. Relative humidity at any temperature is the fraction of the water vapor concentration corresponding to saturation.

Constituent Gas	Dry Air (mol%)	Sat Air @ 100°F (mol%)
Nitrogen	78.09	73.04
Oxygen	20.95	19.60
Water	0.00	6.46
Argon	0.93	0.87
Carbon Dioxide	0.030	0.028
Neon	0.0018	0.0017
Helium	0.000524	0.00049

Antistatic. Having the ability to dissipate charge at a sufficient rate to prevent hazards or nuisances under the conditions of use.

Bonding. The process of connecting two or more conductive objects together by means of a conductor so that they are at the same potential as each other but not necessarily at the same potential as the earth.

Breakdown Strength. The minimum voltage for spark breakdown to occur across a material of given thickness held between electrodes producing a uniform electric field under specified test conditions. Expressed as volts per unit thickness. As the potential across any material is increased, Ohm's law is typically obeyed initially, followed by a region of non-Ohmic behavior; finally a potential is reached at which the current increases at an uncontrolled rate. The corresponding field is the breakdown strength, also known as breakdown field, dielectric strength, and electric strength. Insulator breakdown strengths are typically 10^6 to 10^9 V/m at room temperature, varying considerably with test method. High vacuum has the highest breakdown strength followed by thin homogeneous solids and then liquids.

Breakdown Voltage. The minimum voltage for spark breakdown of a specified gas mixture between electrodes producing a uniform electric field, for specified values of the product of gas density and gap length.

Brush Discharge. A higher energy form of corona discharge characterized by low frequency bursts or streamers. The simplest type forms between charged, isolated nonconductive surfaces and grounded conductors acting as electrodes. For positive electrode, pre-onset or breakdown streamers are observed and the maximum effective energy is 5–10 mJ, while for negative electrode the maximum effective energy is a few tenths of a mJ. Brush discharges may ignite flammable gas and hybrid mixtures but not dust in air. Brush discharges may also form on nonconductive layers with grounded substrates. These include "transitional brush discharges" which have a greater effective energy than brushes from isolated nonconductors but a smaller effective energy than propagating brushes.

Bulking Brush Discharge. A partial surface discharge created during bulking of powder in containers, appearing as a luminous, branched channel flashing radially from the wall toward the center of the container. Its maximum effective energy with respect to dust ignition is believed to be 10–20 mJ (less than the MIE of *Lycopodium*). May ignite flammable gas, hybrid mixtures and some fine dusts in air. Owing to large size of phenomenon the effective energy cannot be measured directly but instead must be inferred from accident case histories where dust of known spark MIE was ignited with proper grounding in place.

Capacitance (*C*). The constant of proportionality between charge and potential difference for systems of conductive bodies. The capacitance in Farads is the charge in Coulombs which must be communicated to raise the potential difference by 1 V.

Charge (Q). An excess or deficit of electrons expressed in Coulombs. An electron carries an electric charge of -1.6×10^{-19} Coulombs.

Charge Density (Q_v). The quantity of charge divided by the total volume that contains this charge, expressed in Coulombs per cubic meter.

Charge-to-Mass Ratio (Q/m). Charge density expressed per unit mass of condensed phase material. Also known as specific charge.

Charge Relaxation. The process by which separated charges recombine or excess charge is lost from a system.

Charging Current (I_c). The rate of flow of charge into a given system per unit time, expressed in Amperes.

Class. The NFPA system for flammable and combustible liquids is given in NFPA 30. Class I flammable liquids have closed-cup flash points below 100°F (37.8°C) and Reid vapor pressures not exceeding 40 psia (2068.6 mmHg) at 100°F (37.8°C). Class I flammable liquids are further classified as follows. Class 1A liquids have flash points below 73°F (22.8°C) and boiling points below 100°F (37.8°C). Class IB liquids have flash points below 73°F (22.8°C) and boiling points at or above 100°F (37.8°C). Class IC liquids have flash points at or above 73°F (22.8°C) but below 100°F (37.8°C). Combustible liquids have closed cup flash points at or above 100°F (37.8°C) and are subdivided into Class II and Class III as follows. Class II liquids have flash points at or above 100°F (37.8C) but below 140°F (60C). Class IIIA liquids have flash points at or above 140°F (60°C) but below 200°F (93°C). Class IIIB liquids have flash points at or above 200°F (93°C). The DOT system for flammable and combustible liquids is given in 49CFR Part 173. Disregarding special cases, flammable liquids have flash points of not more than 60.5°C (141°F) and combustible liquids have flash points above 60.5°C (141°F) and below 93°C (200°F).

Combustible. Capable of burning.

Combustible Dust. (see also "Powder"). Any finely divided solid material, 425 μm or less in diameter (i.e., material passing through a U.S. No. 40 standard sieve), that presents a fire or explosion hazard when dispersed and ignited in air or other gaseous oxidizer.

Conductive. Having a conductivity greater than 10^4 pS/m or a resistivity less than 10^8 Ω-m.

Conductive Floor. Having an average resistance between 2.5×10^3 Ω and 1×10^6 Ω as measured using specified electrodes placed a specified distance apart (Method: NFPA 99).

Conductive Hose. Having an electrical resistance less than 10^3 Ω per meter of hose length based on measurement between the end connectors. Lower resistance values are required in certain cases, for example 6 Ω/m for water washing hoses in ISGOTT.

Conductivity (κ). The reciprocal of resistivity expressed as Siemens per meter or more usually in picosiemens per meter, where 1 picosiemen (pS) = 1×10^{-12} Siemen. Test methods: [143–146].

Cone Discharge. See "Bulking Brush Discharge."

Corona Discharge. A self-sustained low energy electrical discharge with nonthermal ionization that takes place in the vicinity of an electrode of sufficiently low radius of curvature, in a medium whose pressure is typically close to atmospheric. May be accompanied by a hissing noise that increases with current and may be observed as a pinpoint of bluish light at the electrode in darkened surroundings. The ionization region is confined to a small volume close to the electrode while in the remainder of the interelectrode space the ions accumulate and drift due to the electric field without additional ionization, creating ionic wind. Corona discharges are usually observed with electrode radii of curvature less than about 3 mm and especially at points. The effective energy depends on current; ignition of hydrocarbon vapor in air has been reported for currents exceeding 200 μA. Only unusually sensitive gas mixtures such as CS_2, H_2, C_2H_2 in air, or gases in oxygen enriched atmospheres, may be ignited by typical coronas.

Coulomb (C). The quantity of electricity on the positive plate of a capacitor of 1 Farad capacitance when the potential difference across the plates is 1 V.

Coulomb's Law. Force between two charged bodies is proportional to the product of the two charges and inversely proportional to the square of the distance between them.

Current (*I*). The rate of transfer of electricity normally expressed in Amperes (Coulombs per second).

Current Density (*J*). The current (Amperes) passing through unit area of a material (square meters).

Deflagration. Propagation of a combustion zone at a velocity less than the speed of sound in the unreacted medium.

Dielectric Constant (ε_r). A dimensionless parameter expressing the ratio of the permittivity of a material to that of vacuum. Metals have an infinite dielectric constant while gases and vapors have a dielectric constant close to unity.

Dissipation. See "Effective Conductivity."

Effective Conductivity. The charging process in liquids of low conductivity alters the concentration of charge carriers. In the charged state, such as downstream of microfilters, petroleum products with measured (rest) conductivities above about 2 pS/m often display a relaxation time a factor of 3–4 times longer than predicted. However, for rest conductivities less than about 2 pS/m, relaxation times become much shorter than predicted. For example, the relaxation time of jet fuel having a conductivity of 0.01 pS/m is about 30 min based on Ohmic relaxation. In practice it is found that highly charged, nonviscous liquids of such low conductivity relax charge hyperbolically rather than Ohmically. The provision of 100 s of residence time downstream of filters is found to reduce the charge to about 5% of its initial value, whereas via Ohmic relaxation this would require three relaxation times (>1 hour) of residence time. In Appendix B, nonviscous liquids whose rest conductivities are usually measured at about 2 pS/m or less are not assigned a relaxation time but instead a 100 s "dissipation time."

Effective Energy of Static Discharge. The spark ignition energy (*J*) of the least easily ignitable flammable mixture known to have been ignited by the static discharge.

Electric Field Intensity (*E*). A measure of the force exerted by one charged body on another. Imaginary "lines of force" or "electric field lines" originate (by convention) on positive charges and terminate on negative charges. They can be thought of as elastic lines which repel each other in a direction perpendicular to the line itself. The electric field intensity (volts/meter) at any location is the force (Newtons) that would be experienced by unit test charge (Coulombs) placed at the location. A uniform electric field is an ideal case in which the electric field lines are parallel with one another, for example between the plates of a large, parallel plate air capacitor. A divergent electric field is one in which the field intensity changes with distance, for example in a capacitor comprising a sphere and a plate. In practical situations electric fields are rarely uniform, particularly within solids and liquids.

Electrophoresis. The phenomenon of particle migration in an electric field.

Electrophorus. An early device for generating charge by induction, attributed to and named in 1775 by Volta. Comprises a nonconductive plate or "cake" which can be charged by rubbing, and an isolated metal plate which can be placed on the charged cake. After contact is made, a charge is induced on the metal plate; if the top of the metal plate is grounded momentarily, the "like" charge induced by the cake is conducted to ground while an equal and opposite charge remains on the metal plate. After the plate is lifted from the cake, it retains the opposite charge. The process may be repeated indefinitely to produce charge on metal plates without recharging the cake. Charge transfers are typically of the order 1 μC.

Farad. A unit of capacitance equal to Coulombs per volt. See "Capacitance."

Faraday Pail (Cage). Grounded metal sheet or mesh enclosure that prevents external electric fields from exerting an influence inside it.

Flammable Limits. The minimum and maximum molar (or volume) concentrations of a combustible material in a homogeneous mixture with a gaseous oxidizer that will propagate a flame.

Flash Point. The minimum temperature at which a liquid gives off sufficient vapor to form an ignitable mixture with air within the test vessel used (Methods: ASTM 502). The flash point is less than the "fire point" at which the liquid evolves vapor at a sufficient rate for indefinite burning.

Fundamental Burning Velocity (S_u). The burning velocity is the rate of flame propagation relative to unburned gas ahead of the flame front. The fundamental burning velocity is the burning velocity of a laminar flame under stated conditions of composition, temperature and pressure in the unburned gas. The reported S_u is usually the maximum value measured for the fastest burning composition at 25°C and 760 mmHg.

Gallon. US Liquid Measure equal to 3.785×10^{-3} m^3.

Go-Devil. See "Surface Streamer."

Grooved and Shouldered Joint. Type of pipe connection that might result in electrical isolation by gasket and which may require jumper cable across joint.

Ground Indicator. Commercially available device which automatically senses the presence of ground continuity and which may be equipped with alarms or interlocks to prevent operations such as flammable liquid transfers unless equipment is grounded.

Grounding (Earthing). Ideally the process of connecting one or more conductive objects to ground so that each is at the same potential as the earth. By convention the earth has zero potential. In practice, grounding is the process of providing a sufficiently small resistance to ground so that a static hazard cannot be created at the maximum credible charging current to a system.

Hybrid Mixture. A mixture of a flammable gas with either a combustible dust or combustible mist.

Ignitable Mixture. A flammable gas, mist, or dust mixture or any combination of these that can be ignited by a specified ignition source such as a static spark.

Ignition. Self-sustained flame propagation caused by a static discharge in a flammable mixture.

Ignition Energy. Energy in Joules stored in a capacitor which upon discharge is just sufficient to effect ignition of a given fuel mixture under specified test conditions (Methods: ASTM 582 for Gas Mixtures).

Ignition Sensitive. "Easily ignitable" gases or dusts meeting the following criteria: a gas whose LMIE in air is small compared with the 0.2 mJ value typical of common fuel and solvent vapors, or a dust whose MIE in air is small compared with that of *Lycopodium*. In practice, ignition sensitive gases have LMIEs $\leq$ 0.1 mJ and ignition sensitive dusts have MIEs $\leq$ 10 mJ. Examples of ignition sensitive gases are acetylene and hydrogen. Ignition sensitive dusts include finely divided bisphenol-A (BPA) and butylated hydroxy toluene (BHT).

Impedance. Frequency-dependent counterpart of resistance in AC circuits, comprising both inductive and resistive components.

Incendive. Ability of an energy source, such as a static discharge, to cause ignition in a given system.

Induction Charging. This is the process of momentarily grounding a conductor which has been polarized by an electric field, then removing the conductor from the electric field so that it gains a net charge.

Inert Gas. A nonflammable, nonreactive gas that renders the combustible material in a system incapable of supporting combustion.

Insulating. See "nonconductive."

Insulating Flange. A flange or spool piece inserted into an otherwise conductive pipe–hose system. These are not truly insulating but have typical resistances of about 10 kΩ, which is high enough to prevent hazardous stray currents but low enough to prevent hazardous static accumulation. Used particularly in liquid or vapor lines subject to stray current arcs (breaksparks) when hoses are disconnected.

Ionic Wind. In a corona discharge the interelectrode space contains slow moving ions of a single polarity corresponding to the polarity of the sharp electrode. These make frequent collisions with neutral molecules of similar size to the ions resulting in momentum transfer and bulk gas movement in the direction of the electric field. The bulk gas movement, having a velocity of the order 1 m/s, is known as the ionic wind.

Lap Joint. See "Slip Flange."

Laplace Equation. Partial differential equation applying to potential distribution for any system of conductors whose intervening space contains no free charges. Has unique solution for given boundary conditions.

Lightning-like Discharge. Hypothetical discharge in large volumes containing clouds of charged, suspended material (such as silos being loaded with powder and supertanker tanks being water washed). Is observed in nature during dust storms, tornadoes and volcanic eruptions. Owing to limited size and/or charge density in practical containers, such discharges are highly improbable and have not been reported.

Limiting Oxidant (Oxygen) Concentration (LOC). The lowest molar (or volume) gas-phase concentration of oxidant (usually oxygen) at which a specified fuel can propagate a flame.

Lower Flammable Limit (LFL). The lowest molar (or volume) concentration of a combustible substance in an oxidizing medium that will propagate a flame.

***Lycopodium*.** Any of about 200 varieties of club moss. The commonest is the staghorn moss *Lycopodium clavatum* from which a resinous, reticulate-shaped pollen is obtained. This is commercially available and is characterized by a relatively spherical, uniformly sized particle (surface average diameter 29 μm, typically with 80 wt% contained within 25–35 μm) suitable for calibration purposes. When tested by noninductive sparks in various dust ignition energy apparatus under the auspices of ASTM, its minimum ignition energy has been found to be about 20 mJ. This is believed to correspond to, or exceed, the maximum effective energy of bulking brush discharges.

Maurer Discharge. See "Bulking Brush Discharge."

Mil. One thousandth part of one inch, equal to 25.4 μm.

Minimum Explosible Concentration (MEC). The lowest concentration of a combustible dust in air, expressed in grams per cubic meter, that will propagate a flame.

Minimum Ignition Energy (MIE). Energy in Joules stored in a capacitor which upon discharge is just sufficient to effect ignition of a given fuel mixture under specified test conditions. The lowest minimum ignition energy (LMIE) is the MIE of the most easily ignitable composition of the fuel mixture.

Minimum Ignition Voltage. Voltage across capacitor of specified capacitance which upon discharge is just sufficient to effect ignition of the most easily ignitable composition of a given fuel–oxidant mixture.

Mist. A dispersion of fine liquid droplets in a gaseous medium.

Nonconductive. Possessing a conductivity less than 10^2 pS/m or a resistivity greater than 10^{10} Ω-m.

Ohm (Ω). See "Resistance."

Ohm's Law. The Current Density (Amperes per square meter) divided by Electric Field Intensity (volts per meter) is a constant known as Conductivity (Siemens per meter). For small conductors such as wires, the voltage between two points (volts) divided by the current (Amperes) is a constant known as Resistance (Ohm).

Permittivity (ε or $\varepsilon_r \varepsilon_0$). The property of a material determining its ability to store energy when placed in an electric field, equal to the product of dielectric constant and permittivity of vacuum, expressed in Farad per meter. Energy storage is related to the polarizability of molecules in the electric field. Hence polar molecules have higher dielectric constants.

Permittivity of Vacuum (ε_0). A constant equal to 8.854×10^{-12} F/m.

Pico. Subdivision equal to one part in 10^{12}.

Poisson's Equation. A more general form of the Laplace equation in which the intervening space between conductors may contain free charges.

Polarization. The process by which charges are moved to new locations on a conductor by the creation of an electric field acting on the conductor or the movement of the conductor into the influence of an existing electric field. Electron redistribution occurs instantly in metals because owing to their infinite dielectric constants, internal electric fields cannot exist. The term is also used to describe the response of polar molecules to an electric field.

Potential (ϕ or V). The potential (ϕ) any point is measured by the work required to bring unit positive charge from an infinite distance. Potential difference (V) between two points is measured by the work required to carry unit positive charge from one to the other. If the work is expressed in Joules and charge in Coulombs the potential difference is expressed in Volts. In practice, potentials are expressed relative to the earth which by convention is assigned a potential of zero volts. The potential difference between two points is commonly expressed as the difference in their potentials (volts) relative to that of the earth.

Powder. Generic term for subdivided solid material, comprising pellets, granules and dust. Pellets have a diameter greater than 2 mm and typically above 3 mm, although they may contain granules and small quantities of dust. Granules (e.g., granulated sugar) have a diameter between 0.42 and 2 mm, although granular powders typically contain dust. Dusts have a diameter less than 0.42 mm and as low as 1 μm. Suspended particles smaller than 1 μm are known as fumes. In European publications, pellets are typically described as "granules."

Propagating Brush Discharge. An energetic discharge caused by electrical breakdown of the dielectric layer in a capacitor and massive lateral surface discharge resulting in dissipation of most of the stored charge. The capacitor is typically formed by charged plastic coating on a metal substrate although plastic pipe and tote bins may also form the required charged double layer. The stored energy may exceed 1000 mJ, presenting both personnel shock and ignition hazards for a wide variety of materials, including dusts in air.

Relaxation Time (τ). Time constant equal to the time in seconds for charge to decay by Ohm's Law to e^{-1} (36.7%) of its initial value. For a capacitor, the time constant (s) is the product of resistance (Ω) and capacitance (F). For a charged material sample, either solid or liquid, the time constant is the product of permittivity (F/m) and resistivity (Ω-m).

Resistance (*R*). The property of conductors depending on material, dimensions and temperature which determines the current produced by a given difference of potential. The practical unit of resistance, the ohm, is that resistance through which a difference of potential of 1 V will produce a current of 1 A.

Resistivity (ρ_v). The volume resistance of a sample of material having unit length and unit cross-sectional area. If the length is expressed in meters and the area in square meters; the resistivity is given in ohm-meters (Methods ASTM D991-83 and D257-78).

Semiconductive. Possessing a conductivity between 10^2 and 10^4 pS/m or a resistivity between 10^8 and 10^{10} Ω-m.

Semiconductive Hose. Having an electrical resistance great enough to limit the flow of stray electric current to safe levels, yet not so great as to prevent relaxation of static electrical charges to ground. Typically achieved using carbon-loaded layers yielding 10^3 to 10^5 Ω/m of hose length. In a flammable atmosphere the resistance from any metal connector or wand to ground should be less than 10^6 Ω. If only a thin inner layer of a solids transfer hose is semiconductive it should be ensured that abrasion does not cause an unacceptable increase of resistance.

Siemen (S). Reciprocal ohm, formerly termed a "mho." See "Conductivity."

Slip Flange. Loose "ring" flange for connecting flared or stub end pipe. Continuity may be lost across connection due to insulation by nonconductive gasket plus painted surfaces on the slip flange and/or pipe. A conductive gasket (such as flexible graphite filled, spiral wound metal type) or jumper cable may be required.

Spark Discharge. Transient discrete electric discharge which takes place between two conductors which are at different potentials, bridging the gap in the form of a single ionization channel.

Static Electricity. The class of phenomena recognized by the presence of electrical charges, either stationary or moving, and the interaction of these charges, this interaction being solely by reason of the charges themselves and their position and not by reason of their motion.

Stray Capacitance. Capacitance unessential for the function of a circuit and of relatively small value that nevertheless may introduce errors if not eliminated or unaccounted for. Of particular importance in high voltage, low capacitance spark ignition tests and in the measurement of small charges and currents.

Stray Currents. Electric currents which wholly or in part do not follow the original circuits designed for them.

Streaming Current. The charging current introduced to a system by liquid flow, expressed in Amperes.

Surface Charge Density (Q_s or s). The charge per unit area of solid surface expressed in Coulomb/square meter. For a unidirectional, uniform field in air the maximum surface charge density before air breakdown occurs is 2.65×10^{-5} C·m^{-2}. Only a fraction of this value is achieved in most practical situations. However, for nonuniform fields around a charged particle much greater charge densities can be supported in air, depending on the radius of the particle. Also, much greater surface charge densities can be supported when nonconductive charged surfaces have a metal substrate so as to form a capacitor. Since the electric field is primarily exerted through the nonconductor to the substrate the maximum charge density may be limited by the breakdown strength of the nonconductor.

Surface Resistivity (ρ_s). The resistance across a surface element having unit length and unit width, expressed in ohm/square or ohm (Method : ASTM D257). In EOS/ESD S11.11-1993, a static dissipative packaging material is defined as having a surface resistivity equal to or greater than $1.0 \times 10^4\ \Omega$ and less than $1.0 \times 10^{11}\ \Omega$, while an insulative packaging material has a surface resistivity equal to or greater than $1.0 \times 10^{11}\ \Omega$.

Surface Streamer. A wall-to-surface discharge observed on charged liquids during tanker filling, appearing as a flash up to 2 ft long and accompanied by crackling sound. Effective energy unknown but possibly up to about 10 mJ. Also observed on tribocharged solid surfaces such as between aircraft windshields and metal airframe, where puncturing does not occur.

Transitional Brush Discharge. A partial discharge from a nonconductive layer on a conductive substrate. The effective energy is highly dependent on the breakdown voltage of the layer and is intermediate between brush discharges and propagating brush discharges. The more energetic of these discharges might ignite dust in air.

Tribocharging. Charge separation caused by rubbing of surfaces, creating "triboelectricity."

Trillion. The cube of ten thousand. Similarly in the US system, a billion is the cube of one thousand and a million is the cube of one hundred.

Upper Flammable Limit (UFL). The highest molar (or volume) concentration of a combustible substance in an oxidizing medium that will propagate a flame.

Viscous Nonconductive Liquid. Nonconductive liquid having a kinematic viscosity above 30–100 cS and a very slow rate of charge dissipation, equivalent to a conductivity of about 0.02 pS/m and a relaxation time of about 1000 s.

Voltage (V). See "Potential."

Work Function. The minimum energy required to extract the weakest bound electron from a particle surface and remove it to infinity.

Zone. Classification system for electrical and electronic equipment and wiring for all voltages in locations where fire or explosion hazards may exist, as defined in Article 505 of NFPA 70 "National Electrical Code."

INDEX

H

I

L

M

N

O

P

V

W

Publications Available from the
CENTER FOR CHEMICAL PROCESS SAFETY
of the
AMERICAN INSTITUTE OF CHEMICAL ENGINEERS
3 Park Avenue, New York, NY 10016-5901

CCPS Guidelines Series

Guidelines for Pressure Relief and Effluent Handling Systems
Guidelines for Design Solutions for Process Equipment Failures
Guidelines for Safe Warehousing of Chemicals
Guidelines for Postrelease Mitigation in the Chemical Process Industry
Guidelines for Integrating Process Safety Management, Environment, Safety, Health, and Quality
Guidelines for Use of Vapor Cloud Dispersion Models, Second Edition
Guidelines for Evaluating Process Plant Buildings for External Explosions and Fires
Guidelines for Writing Effective Operations and Maintenance Procedures
Guidelines for Chemical Transportation Risk Analysis
Guidelines for Safe Storage and Handling of Reactive Materials
Guidelines for Technical Planning for On-Site Emergencies
Guidelines for Process Safety Documentation
Guidelines for Safe Process Operations and Maintenance
Guidelines for Process Safety Fundamentals in General Plant Operations
Guidelines for Chemical Reactivity Evaluation and Application to Process Design
Tools for Making Acute Risk Decisions with Chemical Process Safety Applications
Guidelines for Preventing Human Error in Process Safety
Guidelines for Evaluating the Characteristics of Vapor Cloud Explosions, Flash Fires, and BLEVEs
Guidelines for Implementing Process Safety Management Systems
Guidelines for Safe Automation of Chemical Processes
Guidelines for Engineering Design for Process Safety
Guidelines for Auditing Process Safety Management Systems
Guidelines for Investigating Chemical Process Incidents
Guidelines for Hazard Evaluation Procedures, Second Edition with Worked Examples
Plant Guidelines for Technical Management of Chemical Process Safety, Revised Edition
Guidelines for Technical Management of Chemical Process Safety
Guidelines for Chemical Process Quantitative Risk Analysis
Guidelines for Process Equipment Reliability Data with Data Tables
Guidelines for Safe Storage and Handling of High Toxic Hazard Materials
Guidelines for Vapor Release Mitigation

CCPS Concepts Series

Avoiding Static Ignition Hazards in Chemical Operations
Estimating the Flammable Mass of a Vapor Cloud
RELEASE: A Model with Data to Predict Aerosol Rainout in Accidental Releases
Local Emergency Planning Committee Guidebook: Understanding the EPA Risk Management Program Rule
Inherently Safer Chemical Processes. A Life-Cycle Approach
Contractor and Client Relations to Assure Process Safety
Understanding Atmospheric Dispersion of Accidental Releases
Expert Systems in Process Safety
Concentration Fluctuations and Averaging Time in Vapor Clouds

Proceedings and Other Publications

Proceedings of the International Conference and Workshop on Reliability and Risk Management, 1998
Proceedings of the International Conference and Workshop on Risk Analysis in Process Safety, 1997
Proceedings of the International Conference and Workshop on Process Safety Management and Inherently Safer Processes, 1996
Proceedings of the International Conference and Workshop on Modeling and Mitigating the Consequences of Accidental Releases of Hazardous Materials, 1995
Proceedings of the International Symposium and Workshop on Safe Chemical Process Automation, 1994
Proceedings of the International Process Safety Management Conference and Workshop, 1993
Proceedings of the International Conference on Hazard Identification and Risk Analysis, Human Factors, and Human Reliability in Process Safety, 1992
Proceedings of the International Conference and Workshop on Modeling and Mitigating the Consequences of Accidental Releases of Hazardous Materials, 1991
Safety, Health and Loss Prevention in Chemical Processes: Problems for Undergraduate Engineering Curricula

Printed and bound by CPI Group (UK) Ltd, Croydon, CR0 4YY
16/09/2026
14963420-0004